The Patrick Moore Practical Astronomy Series

The Constellation Observing Atlas

Grant Privett and Kevin Jones

Illustrated by Kevin Jones

 Springer

Grant Privett
Salisbury
Wiltshire
United Kingdom

Kevin Jones
Calne
Wiltshire
United Kingdom

ISBN 978-1-4614-7647-4   ISBN 978-1-4614-7648-1   (eBook)
DOI 10.1007/978-1-4614-7648-1
Springer New York Heidelberg Dordrecht London

Library of Congress Control Number: 2013944210

Cover illustration: Kris Black Design

Printed on acid-free paper

Springer is part of Springer Science+Business Media (www.springer.com)

Acknowledgements

The authors would like to thank Wendy Jones and Rachel Privett; for putting up with them while this book was written, and for the coffee and proof reading. We promise it won't happen again. Well, probably not for a while, anyway.

We would also like to extend our thanks to John Watson, for his encouragement during the preparation of this book and to Philip Moore, who provided so much help and support in creating the maps, and who was so patient when we got it wrong.

Thanks are also due to the amateur astronomers Bill Snyder, Chris Picking, David Ratledge, Martin Pugh and Prof. Greg Parker, who so kindly provided most of the fine images showing deep sky objects. All of the deep sky objects featured were selected as observable using normal home equipment – rather than the Hubble Space Telescope or a semi-professional observatory. Details of their personal websites may be found within Appendix 1.

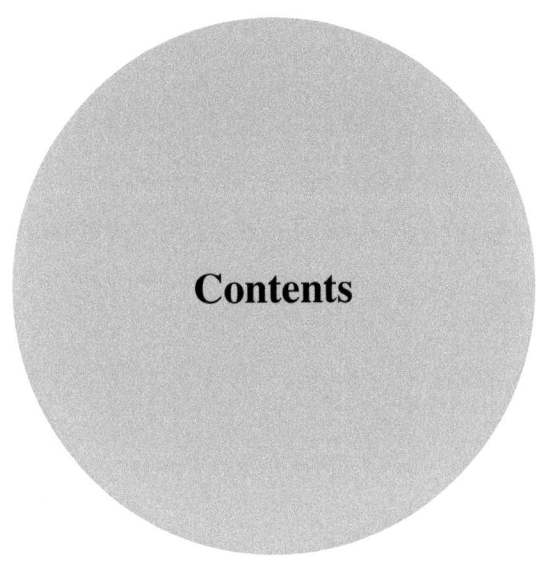

Contents

Introduction: Background and How to Use This Book

RECOGNIZING THE CONSTELLATIONS

If you go out to a dark site on a moonless clear night, at any given moment you will be able to see between 2,000 and 3,000 stars. The brightest of these form the shapes and patterns we recognize as the constellations. It would be daunting and pointless to try to remember the locations of all these stars. But to find your way around the sky and hunt down attractive double stars, variable stars or deep sky objects, you need only memorize the constellation outlines to provide a sky-wide series of reference points.

Fortunately, humans are very good at pattern spotting and have been finding patterns within the stars for thousands of years – probably as far back as the time of the Great Rift Valley in Africa. Many of the constellation patterns we use today were bequeathed to us by the Babylonians, Greeks and Romans and so come complete with a rich, and sometimes faintly bizarre, associated mythology. Later, additional constellations and many star names were supplied by Arabian astronomers and, more recently, by European astronomers who visited the southern hemisphere during the seventeenth and eighteenth centuries.

However, while a few constellation outlines truly look like their namesakes – Orion the Hunter does look rather like a man with a bow – most do not. Some display so little resemblance that you have to assume that the Greeks were either incredibly imaginative or were drinking far too much when they assigned the names. The abstract patterns led to a plethora of constellations being proposed, many of which grouped together a handful of dim stars that had not been integrated into larger constellations. Constellations with wonderful names such as *Officina Typographica* and *Turdus Solatrius* abounded until order was finally brought to what had been a mapping free-for-all.

The patterns employed by astronomers today were formalized in 1930 when the International Astronomical Union (IAU) assigned stars to 88 constellations, formalized the breakup of the enormous and unwieldy Argo Navis constellation into more manageable bits, and fixed the constellation border outlines for good. Some stars also ended up being shared. In effect, the IAU pinned down the constellation outlines like country borders on a map of Earth and defined reference points for the celestial equivalent of latitude and longitude – declination and right ascension, respectively. It certainly made life easier for everyone involved. At last you knew that if you mentioned a star that was,

G. Privett and K. Jones, *The Constellation Observing Atlas*, The Patrick Moore Practical Astronomy Series, DOI 10.1007/978-1-4614-7648-1_1, © Springer Science+Business Media New York 2013

for example, the second brightest in the constellation of Leo, everyone knew which star you were talking about as the ambiguity had been removed.

Some other patterns – such as the Summer Triangle formed by the bright stars Altair, Deneb and Vega – are also well known and recognized by most amateur astronomers, but these are not constellations and will mean little to most professional astronomers.

Although the stars we see are all moving in orbits around the center of the galaxy, the amount they appear to move across the sky every year – their proper motion – is so tiny it was difficult to measure until recently. It will be thousands of years before the constellations change shape enough for it to become obvious to the casual observer. So, unless medicine makes some pretty stunning breakthroughs, for our purposes the star positions given in the charts within this book will not change significantly during your lifetime.

MAKEUP OF STARS AND GALAXIES

All the stars you see at night are suns similar to our own but have been reduced to mere pinpricks of light by their huge distance from us. Some of the stars we see are smaller than our Sun and will shine on long after our Sun is reduced to a dim smoldering ember; other, more massive stars will live fast, brilliant lives and die young – often in explosions that will briefly outshine the whole galaxy. Our Sun is already five billion years old and is likely to burn for at least another five billion years before entering its giant stage. Then it will shrug off its outer layers to form a short-lived planetary nebula before eventually being reduced to a slowly fading white dwarf star.

The stars you can see are members of our galaxy – a vast stellar conglomeration containing roughly 300 billion suns – together with clouds of dust and gas all bound together by their mutual gravitational attraction. As has been said elsewhere, viewed sideways on, our galaxy would look rather like two fried eggs slapped back to back. Whereas, if viewed from above the stars would take up a spiral pattern similar to that seen around a plughole as water leaves a sink. The spiral is not quite pure and our galaxy is actually a slightly barred spiral, but the edge-on profile remains broadly that of the fried eggs' analogy.

Fig. 1 The Milky Way is a splendid sight on a clear and dark night and provides many opportunities for browsing with binoculars. This view was captured from Australian suburbia and includes Alpha and Beta Centauri, Crux and part of Carina (Image by Grant Privett)

Our Sun lies out toward one edge of the galaxy, and most of the stars you see with the naked eye lie within a 1,000 light year radius and so encompass less than 1 % of the width of our galaxy's spiral disc. When we look out at night we see the individual stars within our immediate neighborhood strewn across our sky. Beyond this, the dim glimmers of light arriving from more distant stars within the disc combine to form a hazy band of light that encircles the sky and forms the Milky Way – a familiar and magical sight in its own right when viewed from a dark location. Against it are seen darkened areas where the dust and gas leftover from when the stars formed still lies, obscuring the light from the stars beyond.

REFERENCE SYSTEM

Because Earth rotates upon its axis, the stars, when viewed from a location north of the equator, appear to rotate about a point in the northern sky – the North Celestial Pole (NCP) – during the course of a night. Similarly, for observers in the southern hemisphere stars appear to rotate about a point in the southern sky – the South Celestial Pole (SCP), while an observer at the equator gets to see the whole sky rotate about two points on opposite horizons.

The poles are used as reference points in astronomy, rather like the North and South poles on Earth. The NCP is defined as +90° declination while the SCP is defined as −90°. The stars halfway between – those passing overhead at the equator – have a declination of 0° and are located on the celestial equator.

A declination value is analogous to the more familiar latitude used on Earth. Similarly, a longitude analog called right ascension is employed to define how far east of a reference point on the celestial equator a location is. The reference point employed to define the zero point of right ascension is the location in the sky where the Sun moves from the southern sky to the northern sky as it makes its circuit during the year. While declination is expressed as degrees, right ascension will normally be found in hours (each hour covering 15° and 24 h covering the whole 360° swathe of the celestial equator).

THE PLANETS

Due to Earth's orbital motion, the Sun appears to travel across the sky passing along a route known as the ecliptic – spending half the year north of the celestial equator and half the year south of it. The planets also appear to closely follow the ecliptic, as does the Moon, give or take a few degrees. The constellations through which the ecliptic runs are known as the zodiac and were named thousands of years ago by the Greeks and probably their Babylonian or Sumerian predecessors, too.

Because the planets are in continual motion, there is no point plotting them on these maps. Instead, websites such as www.heavens-above.com or software such as *The Sky* or *Redshift* should be consulted for the positions of planets on a given night. Knowing the locations of the planets will undoubtedly save you lots of anguish when first learning those constellations through which the ecliptic passes. But, as the sky becomes more familiar territory, you will quickly come to know roughly where to find the major planets.

It's worth remembering that the planets can briefly stray into constellations such as Ophiuchus, a constellation between Scorpius and Sagittarius. Don't look for it in horoscopes. Astrologers clearly didn't want 13 signs of the zodiac and defined the zodiac arbitrarily to avoid it!

If you see a bright star in a constellation and it's on the ecliptic, it is probably a planet rather than a supernova. It's a mistake many observers make, sometimes more than once.

DEEP SKY OBJECTS

Besides stars and planets the constellations contain a number of other denizens – the deep sky objects. These are generally fairly dim and only a dozen or so are visible to the naked eye, even from a dark location. However, when binoculars or a small telescope are used, this number swells to several hundred: the best and most accessible of which are plotted on our maps. Some iconic objects such as the Veil Nebula or Pelican Nebula are not shown, as for most observers, especially those in suburban settings, seeing them would be exceptionally difficult.

When a telescope is used, many of these fuzzy patches of haze resolve into compact balls of thousands of ancient stars – globular clusters – or wider groupings of a few dozen younger stars – open clusters.

Some, however, will not resolve into stars because they are clouds of gas and dust – nebulae – that either reflect the light of nearby stars or emit a dim red or green glow of their own. Many dying stars shrug off shells of gas that for a few thousands of years emit greenish light. These are planetary nebulae. Those that glow a very dim red are generally clouds of gas from which stars might one day form.

The remainder of the unresolved objects are distant galaxies. These are akin to our own galaxy but very remote and can only be resolved into their billions of individual stars by the larger professional observatories or by careful imaging with amateur telescopes.

GUIDE TO THE MAP PAGES

Each constellation is covered over between two or three pages. The first page describes the mythology of the constellation and also discusses the various objects of interest within it. These include the notable variable or double stars, meteor showers and deep sky objects that appear within its bounds. The items shown, and indeed the whole deep sky database employed, has been carefully chosen to favor those that will be visible with a telescope of an aperture 150 mm or so from a suburban location, or an 80 mm from a dark location. In addition, the description also discusses the origin of the constellation name and the sky context – whether northern or southern – or if it is likely to include planets.

On the other page you will find a map of the constellation. It contains the location in the sky and a grid indicating the right ascension and declination. The constellation boundary is denoted by the dashed line.

The size of a star indicates its brightness. The bigger a star is drawn, the brighter it appears in the sky (a key precedes the map pages). The color used indicates its spectral class, with the colors exaggerated somewhat to make them easier to distinguish. A chart is provided as part of the key to indicate the relationship between the color used and the star type. Remember when using them that colors of stars are quite subtle and the map colors are chosen to allow stars of different types to be differentiated.

The dimmest stars seen on the maps are normally 8th magnitude. It should be remembered that the brightest stars are of the order of magnitude 0 – like the brilliant blue white Vega – while the dimmest you can see under dark skies will normally be of the order of magnitude 6, which indicates a star some 250 times fainter. The scale extends downward, with 10×50 binoculars revealing 9th magnitude stars, 300 mm aperture telescopes 13th magnitude stars, while the Hubble Space Telescope has detected stars nearing magnitude 30.

Another key indicates the different types of deep sky object you may find within the maps. Nebulae, globular clusters, open clusters and planetary nebulae are all assigned their own symbols. Many of these are identified on the maps by a number preceded by a few letters. These indicate the catalog they are taken from. For example: M78 is the 78th member of the catalog created by Charles Messier. A full list is:

M	Messier Catalog, by Charles Messier
N	New General Catalog, by John Dreyer
I	Index Catalog of Nebulae and Clusters of Stars, by John Dreyer
Stock	Stock open cluster list, by J. Stock
Mel	Melotte Catalog of Deep Sky Objects, by Philibert Melotte
Cr	Catalog of Open Star Clusters, by Per Arne Collinder
B	List of Dark Nebulae, by E. E. Barnard
Sh2	Sharpless Catalog of HII Regions, by Stewart Sharpless

None of these catalogs are shown in full; they all contain objects that are too dim, too small or just plain non-existent. Those 750 or so objects shown in the book are within range of telescopes of 200 mm aperture, with the majority visible in 150 mm apertures and a goodly proportion accessible to an 80 mm refractor. The choices we have made are arbitrary. We are sorry if we missed out any bright favorites. Most, but not all, constellations contain something of interest to the deep sky observer.

The positions of the stars shown on the map are all taken from the Hipparcos and Yale Bright Star catalogs and are of high accuracy. Details are given in the text, describing some of the more notable stars, accessible double stars or obviously variable, for example. An effort has been made to ensure those discussed are generally readily located and quite bright – and thus easier to track down and observe.

Finally, in those constellations that the Milky Way – our home galaxy – passes through, its presence is indicated via a darker blue color. The boundary chosen may differ somewhat from what you see, as it will depend strongly upon how dark your sky is and how well your eyes are "dark adapted." From a city the Milky Way will be barely visible at all, while from the countryside it is a beautiful sight that becomes more striking as your period away from bright lighting grows.

GREEK ALPHABET

Many bright stars are identified by a Greek letter, i.e., alpha, beta, gamma, etc., followed by the constellation name. These designations were used to indicate those stars that were brightest in a constellation – the brightest being denoted alpha, the second brightest, beta and so on.

However, the observations used were not always accurate, and the stars assigned to each constellation have changed over the years, so that some lack a number of stars. To confuse things further there are instances where deep sky objects, such as omega Centauri (visible to the naked eye), have been assigned a letter.

On the maps herein, the symbol for the Greek letter is used rather than the full name, to avoid clutter. A table identifying each symbol is provided below.

MAP KEY

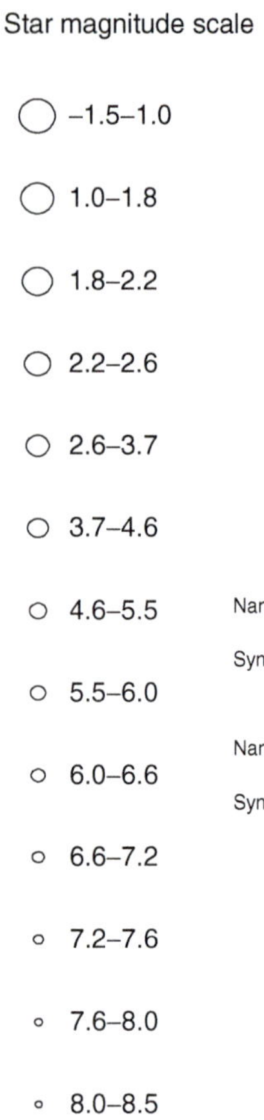

Star magnitude scale

○	−1.5–1.0
○	1.0–1.8
○	1.8–2.2
○	2.2–2.6
○	2.6–3.7
○	3.7–4.6
○	4.6–5.5
○	5.5–6.0
○	6.0–6.6
○	6.6–7.2
○	7.2–7.6
○	7.6–8.0
○	8.0–8.5

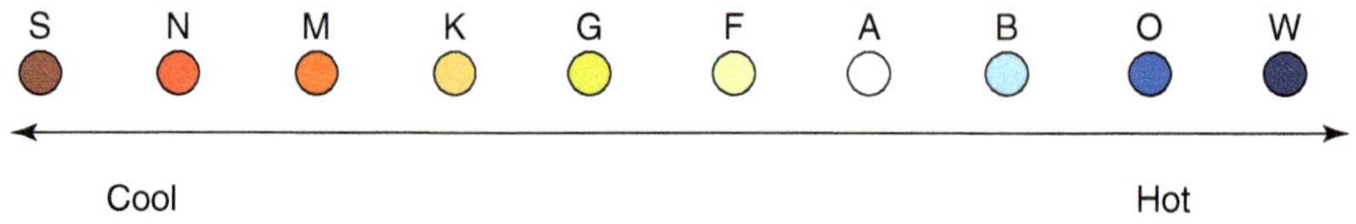

Star spectral type

S N M K G F A B O W

Cool ←————————————————————→ Hot

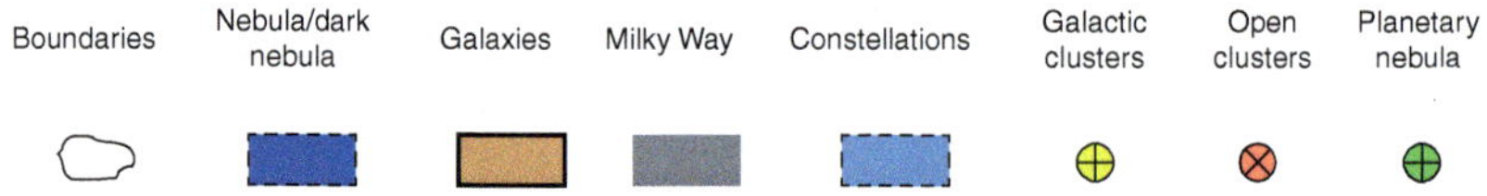

Boundaries　Nebula/dark nebula　Galaxies　Milky Way　Constellations　Galactic clusters　Open clusters　Planetary nebula

Symbols table

Name	Alpha	Beta	Gamma	Delta	Epsilon	Zeta	Eta	Theta	Iota	Kappa	Lambda	Mu
Symbol	α	β	γ	δ	ε	ζ	η	θ	ι	κ	λ	μ

Name	Nu	Xi	Omicron	Pi	Rho	Sigma	Tau	Upsilon	Phi	Chi	Psi	Omega
Symbol	ν	ξ	ο	π	ρ	σ	τ	U	φ	χ	ψ	ω

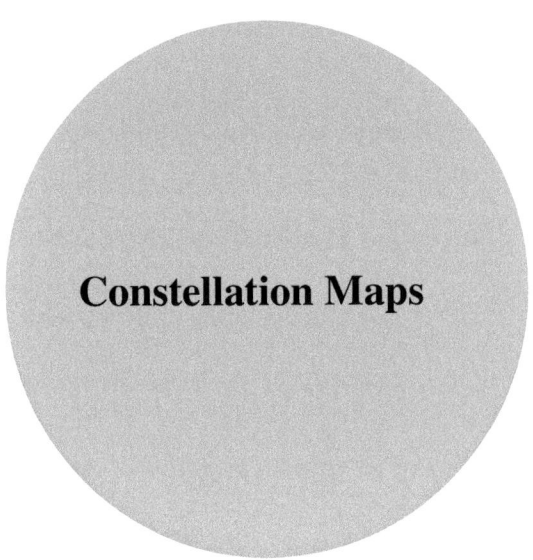

Constellation Maps

ANDROMEDA

A northern hemisphere constellation running eastward from the Square of Pegasus. One of the stars of the Square of Pegasus has now been assigned to Andromeda and is its brightest star. Andromeda has four brightish stars that make it easy to locate. The eastern end of the constellation is quite near the Milky Way; despite this, the area still boasts two of the brightest galaxies visible in the northern sky.

Historically

Andromeda is part of the Perseus quest legend. Andromeda, a fair maid, is tied to a rock to placate Poseidon, who has set about flooding the kingdom ruled by her parents. Poseidon is a bit miffed that Cassiopeia, Andromeda's mother, has said that Andromeda is prettier than Poseidon's daughters, the Nereid sea nymphs. Fortunately, Perseus flies in on his horse, rescues her, and they fly away together. The legend does not relate how Poseidon takes the news: probably, not phlegmatically.

Notable Stars

Double: Gamma Andromedae, 2.3/5.5 and 9.8″. Orange/yellow and blue stars. The blue star may appear greenish due to an optical illusion.

Variable: R Andromedae, a red Mira type star varying semi-regularly between magnitude 6 and 15 as it pulsates.

G. Privett and K. Jones, *The Constellation Observing Atlas*, The Patrick Moore Practical Astronomy Series, DOI 10.1007/978-1-4614-7648-1_2, © Springer Science+Business Media New York 2013

Deep Sky Objects

M31: The Andromeda Galaxy. Visible to the naked eye next to nu Andromedae as a 4th magnitude elongated blur. When seen through a 75mm aperture it becomes more obvious, and its satellite galaxies, M32 and M110, can easily be located, flanking it. M32 is the more compact and brighter of the two. Long exposure images and larger telescopes bring out the spiral structure and dust lanes within this spectacular galaxy. Currently it's some 2.2 million light years away, but will one day collide with the Milky Way. One of the must-see objects in the sky.

NGC7662: A planetary nebula seen as an oval of light some 13 arc seconds across with a central star of magnitude 12–13 – difficult in small instruments. Images show a bright feature within the nebula resembling the Greek letter omega. It should be readily detected as an out-of-focus star in a 114mm reflector. It lies 1,800 light years away.

NGC752: An open cluster on the southeastern edge of Andromeda. It's roughly a degree across and best seen in a small telescope rather than binoculars. Its 5th magnitude, with a brightest star of 9th magnitude.

NGC7686: In the north west of the constellation is a compact 5th magnitude binocular cluster. It contains 40 or more stars.

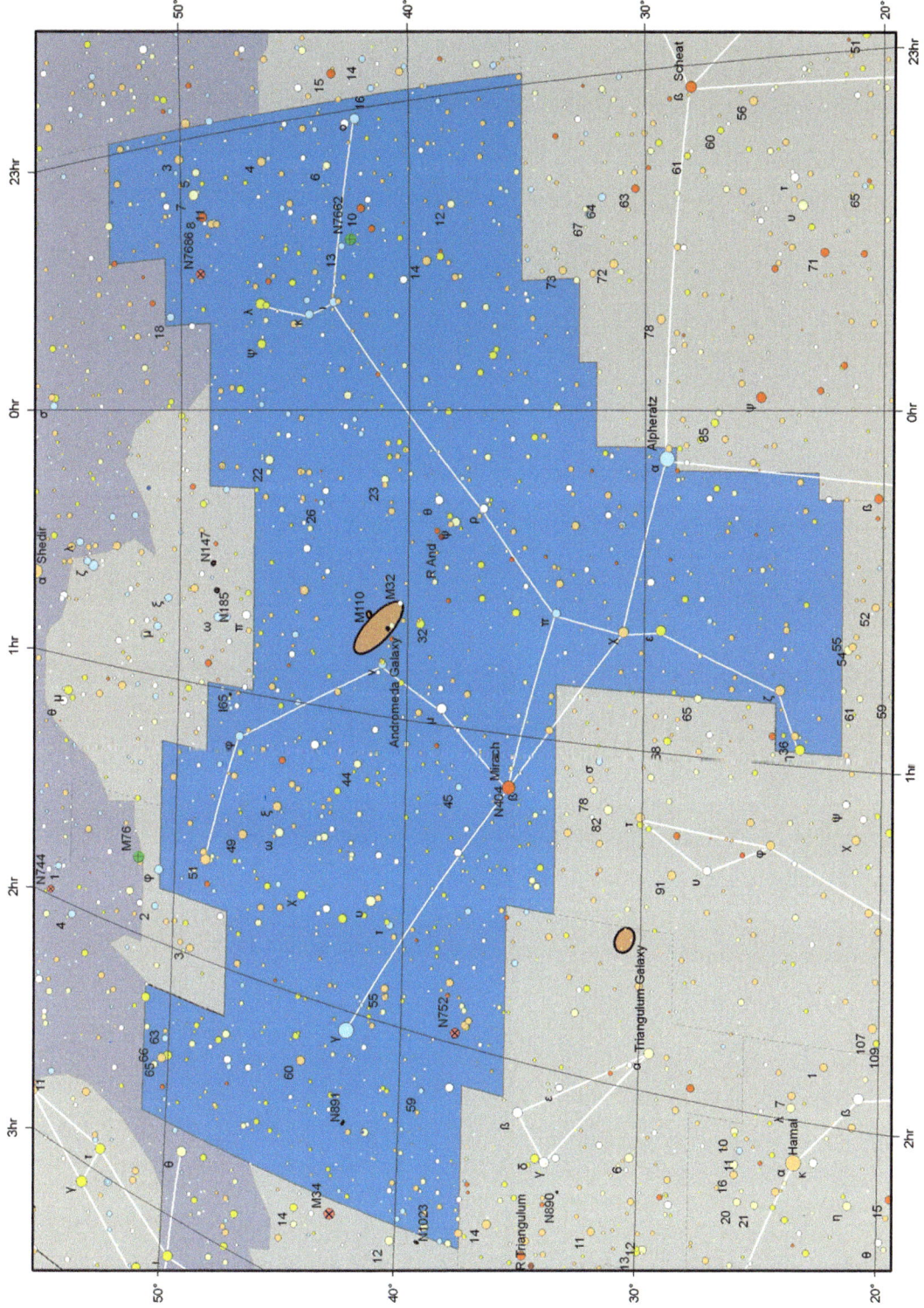

Fig. 1 The superb Andromeda galaxy, obvious to the naked eye and a fine sight in any instrument (Image by Grant Privett)

ANTLIA

A small and rather unimpressive southern hemisphere constellation of 4th and 5th magnitude stars. It is located just north of Vela and so readily found, but there isn't a huge amount to see. Look for it on dark nights during the month of April.

Historically

Antlia is one of the modern constellations created by Europeans travelling south of the equator. The constellation was first described by Nicolas Louis de Lacaille who spent a couple of years at the Cape of Good Hope on the southern tip of Africa (in what is now South Africa) mapping stars. It represents the air pump invented by Denis Papin – who later worked with Robert Boyle.

Notable Stars

Double: Zeta Antliae, 6/6.5 and 8″. White stars.

Double: Delta Antliae, 5.5/9.5 and 11″. Primary is very blue.

Deep Sky Objects

No deep sky objects of note.

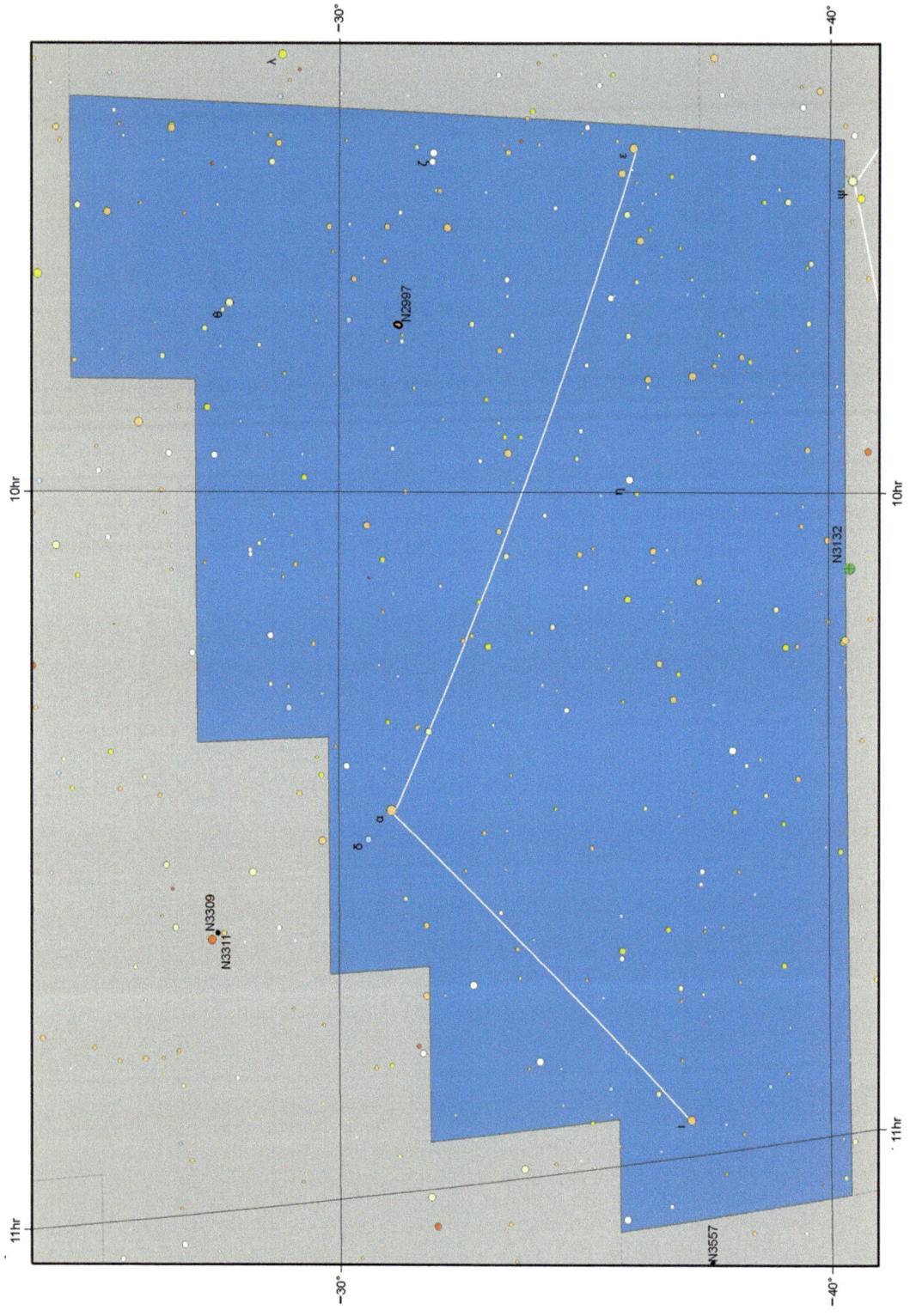

APUS

A southern hemisphere constellation lying quite close to Octans and, hence, close to the South Celestial Pole. It is made up of stars within the magnitude range 3rd to 5th and is little more than two small groupings of stars.

Historically

Apus, a bird of paradise from New Guinea, appeared first on a chart created by Petrus Plancius in 1598. It's not a very memorable constellation: its brightest star magnitude is a magnitude 3.8. There appears to be no associated mythology – although an account of the voyage of Keyser and de Houtman who made the observations that Plancius based his chart on is worth seeking out and reading about.

Notable Stars

Double: Delta Apodis, 4.7/5.3 and 103″. Orange stars. Line of sight double. Suitable for binoculars.

Variable: Theta Apodis is a red M7 star varying semi-regularly between magnitude 5.5 and 6.0 in around 120 days.

Deep Sky Objects

NGC6101: A globular cluster in the northern part of the constellation. It is of 9th magnitude and more than 6 arc minutes across. The individual stars start at magnitude 13 so expect a fuzzy unresolved blob in small telescopes. It contains a surprising number of blue stragglers – stars formed from star mergers.

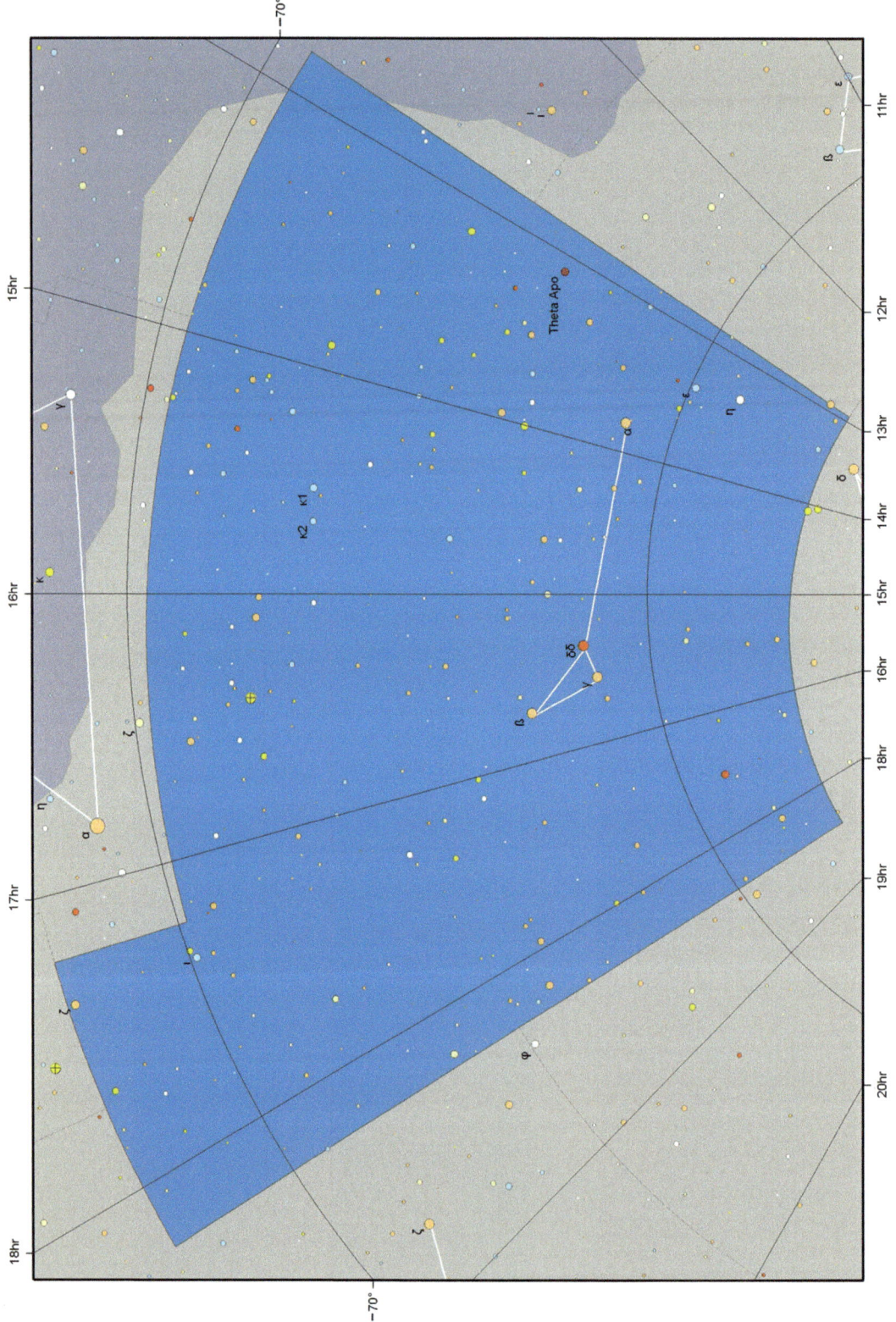

AQUARIUS

A constellation encompassing part of the celestial equator, with the majority of its' 980 square degrees lying in the southern hemisphere. The ecliptic also passes through the constellation. Aquarius is big enough and bright enough to be reasonably easily spotted – especially the group of four stars just east of alpha Aquarii.

The eta Aquarid meteor shower peaks somewhere around May 5. It is best seen from the southern hemisphere. Later in the year, towards the end of July and the start of August, a number of small low-activity radiants appear, providing activity for at least a month.

Historically

A historic constellation, recorded as far back as the Babylonian empire. It depicts a man leaning to pour water, wine or nectar from a vessel. Inevitably, with such an old constellation, various legends account for this character, depending upon the culture considered. Unfortunately, the details of the Babylonian legends are mainly lost.

Those recorded from Greece tend to revolve round Ganymede, a young man, who was by all accounts good looking. Zeus, who turns up in a lot of these legends – rarely honorably – stole Ganymede away to Olympus. Alternatively, the Goddess Eos had Ganymede as a plaything before Zeus stole him away to fill the cups and act as a waiter. Ganymede doesn't seem to have much of a say in things, being a mere mortal. Plots don't get much earthier than those derived from Greek mythology – classical soaps.

Notable Stars

Double: Zeta Aquarii, 4.5/4.5 and 2″. Close double of equally bright F-type white stars.

Double: Beta Aquarii, 3.0/11 and 35″. Wide but very different brightness.

Deep Sky Objects

NGC7293: The Helix nebula is a spectacular and large planetary nebula lying not far from 47 Aquarii in the southern part of Aquarius. It is visible, even in murky skies, using just 50mm aperture binoculars. Through a low power eyepiece on an 80mm aperture it will appear to cover 9 arc minutes of sky, although recent HST images show an outer, dimmer ring twice the width.

NGC7009: Another planetary nebula, but rather smaller than The Helix. It is, however, quite bright and compact at 25×45 arc seconds across and 8th magnitude. It was named the "Saturn" nebula by Lord Rosse. The globular cluster M72 lies nearby and is well worth a look.

M2: In the northern part of the constellation is a 6th magnitude globular cluster containing 150,000 stars. Its brightest individual stars are magnitude 13.

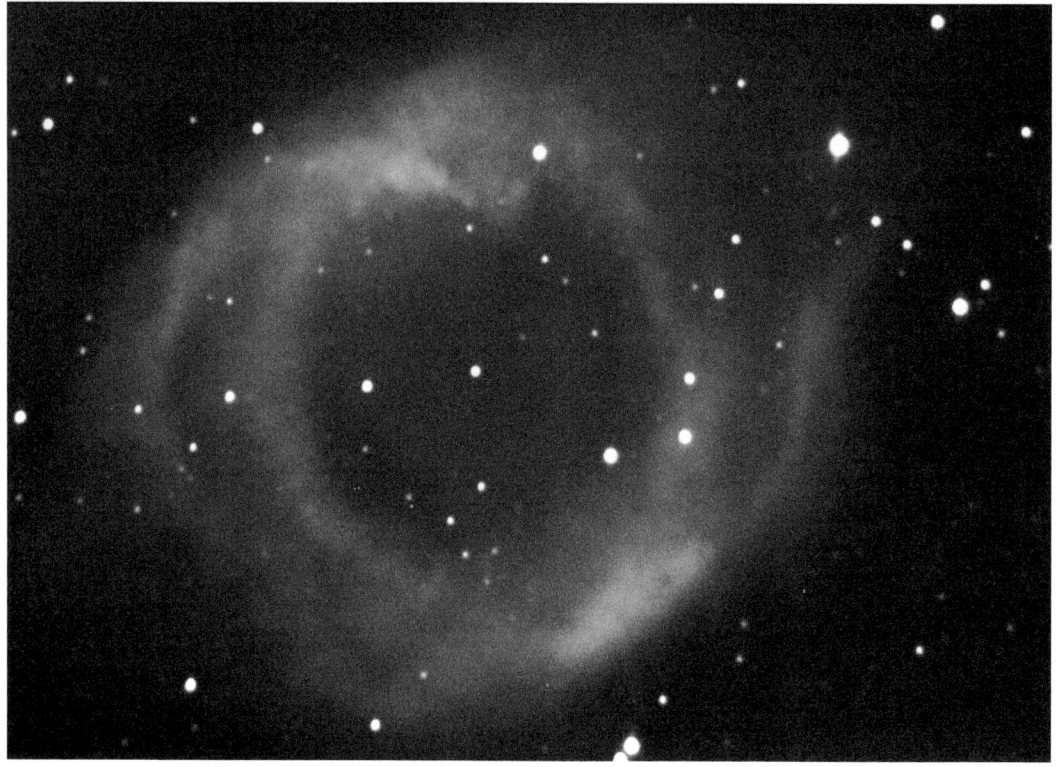

Fig. 2 The Helix nebula is a planetary nebula in Aquarius that is visible in binoculars (Image by Grant Privett)

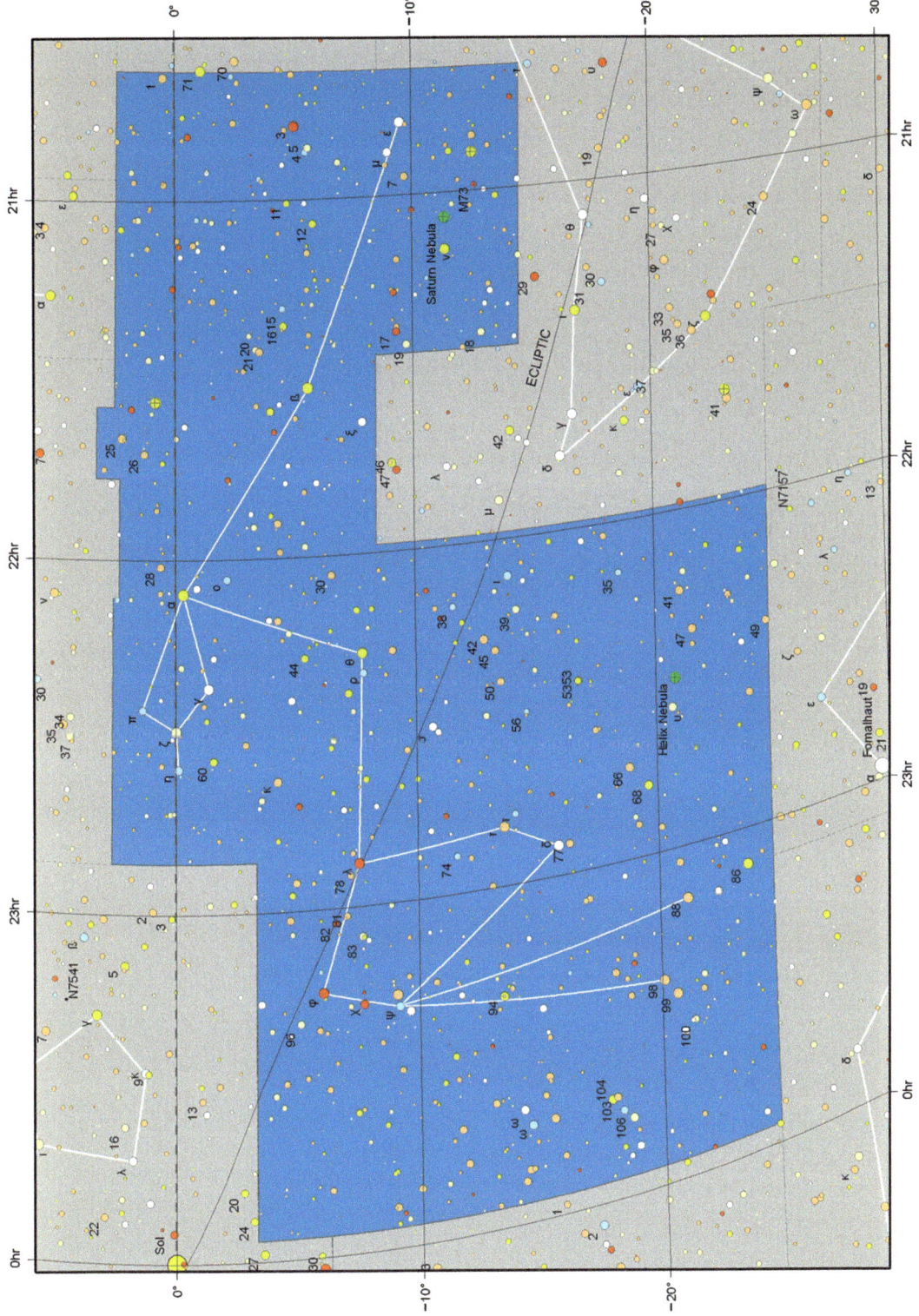

AQUILA

A constellation that bestrides the celestial equator, and is dominated by the grouping of the brilliant star Altair and its companions beta and gamma Aquilae. Altair is one of the three bright stars that make up the "Summer Triangle", as it, Vega and Deneb are known in the northern hemisphere.

Aquila is seen against the backdrop of the Milky Way. The rift that splits the Milky Way in Cygnus runs through the north east side of Aquila and is fantastically detailed when seen in dark, unpolluted skies.

Historically

An old constellation mapped by Ptolemy and, apparently, by Eudoxus of Greece in the 4th century BC – probably based on a Babylonian source. The Romans dubbed it a vulture and borrowed some of its stars to form another long defunct constellation, Antinous. The Greeks managed to relate it to Prometheus, who was tormented by an eagle which disemboweled him every day.

Notable Stars

Double: 57 Aquilae 5.8/6.5 and 36″. Pale yellow and bluish stars. Easily separated with the smallest telescope.

Variable: Eta Aquilae is a Cepheid variable star identified in 1784 by Nathaniel Pigott. It varies between magnitude 3.4 and 4.4 in 7 days.

Deep Sky Objects

NGC6709: A loose scattering of 50 or so stars across 15 arc minutes. The aggregated magnitude is 6th magnitude. Despite lying in a quite rich region of the Milky Way it is reasonably easy to spot with an 80mm refractor, although resolution may require a 130mm aperture.

Milky Way: Despite its location deep in the Milky Way, Aquila is quite poorly supplied with deep-sky objects. There are a number of worthwhile targets, such as the planetary nebulae *NGC6781* (11th magnitude and 1.8 arc minutes across), but they are probably too dim for most users of smaller telescopes. However, the Milky Way itself is well worth patient sweeping with 10×50 binoculars or a wide field telescope. The area contains much of interest, including the dark rift running down through the constellation and into Scutum. From a dark site look for dark nebulae north west of Altair.

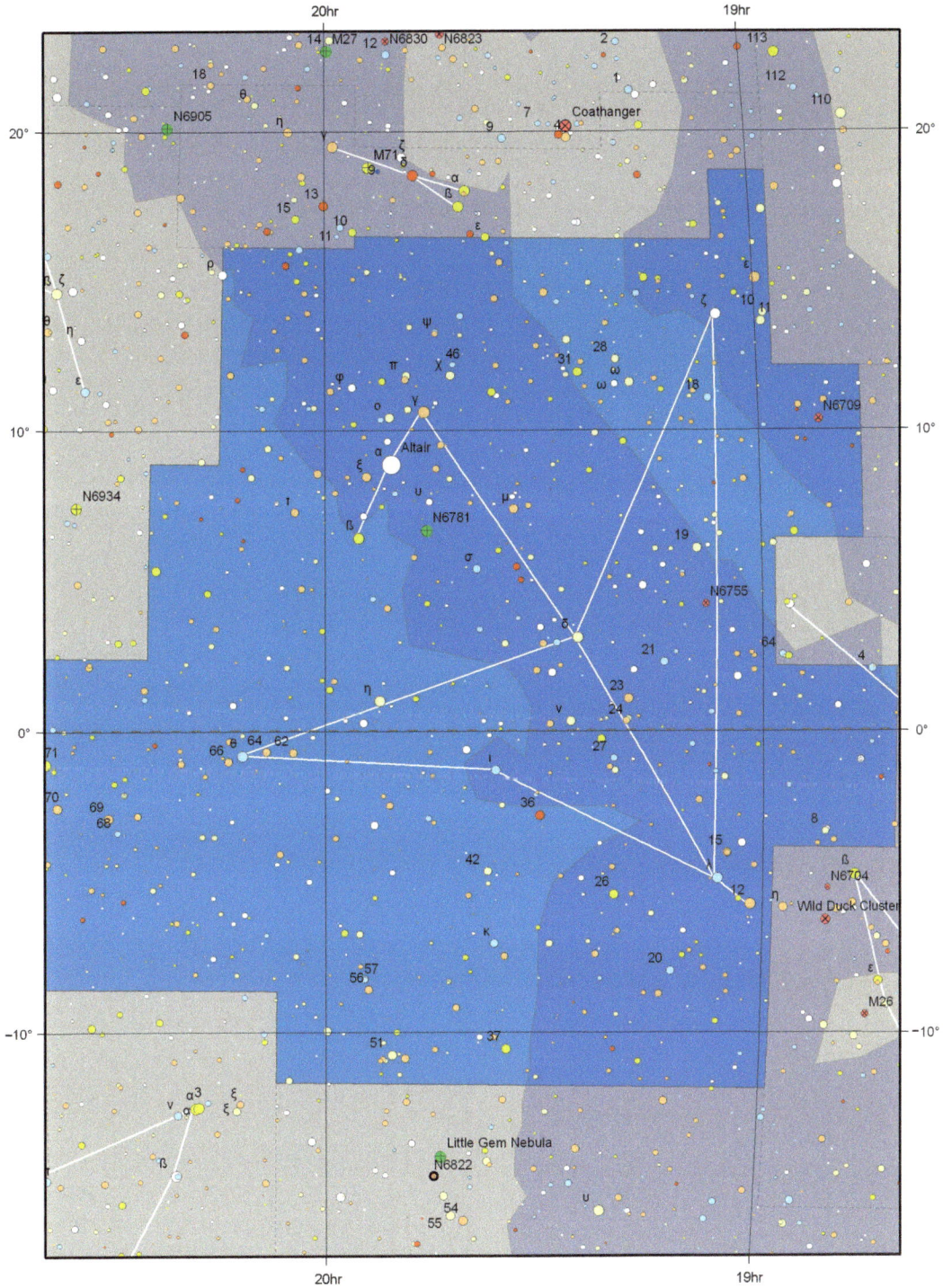

ARA

A small southern hemisphere constellation of 2nd and 3rd magnitude stars lying just south of the easier-to-locate Scorpius and well within the Milky Way. Despite its size it is quite well provided with deep sky objects and very well worth tracking down when at its best: around July.

Historically

The stars of Ara depict an altar. These were a popular feature of older religions, particularly those given to blood sacrifice – gods are rarely vegetarians, it seems. But why place one in the sky? Well, it may have been involved in the clash between the Titans and Greek Gods, but who now can tell for sure?

Notable Stars

Double: Gamma Arae 3.5/10.0 and 18″.

Deep Sky Objects

NC6397: A bright globular cluster of nearly 500,000 stars, with an integrated magnitude of 6.7 and an overall width that of the Moon. It is also one of the nearest, at just over 7,000 light years away, with an age in excess of 13 billion years – nearly as long as the lifetime of the Universe.

NGC6193: A young 5th magnitude open cluster adjacent to a patch of nebulosity. It contains 28 stars within a patch of sky about half the width of the full Moon. The area contains a number of young intense blue stars as part of an OB association centered on the cluster.

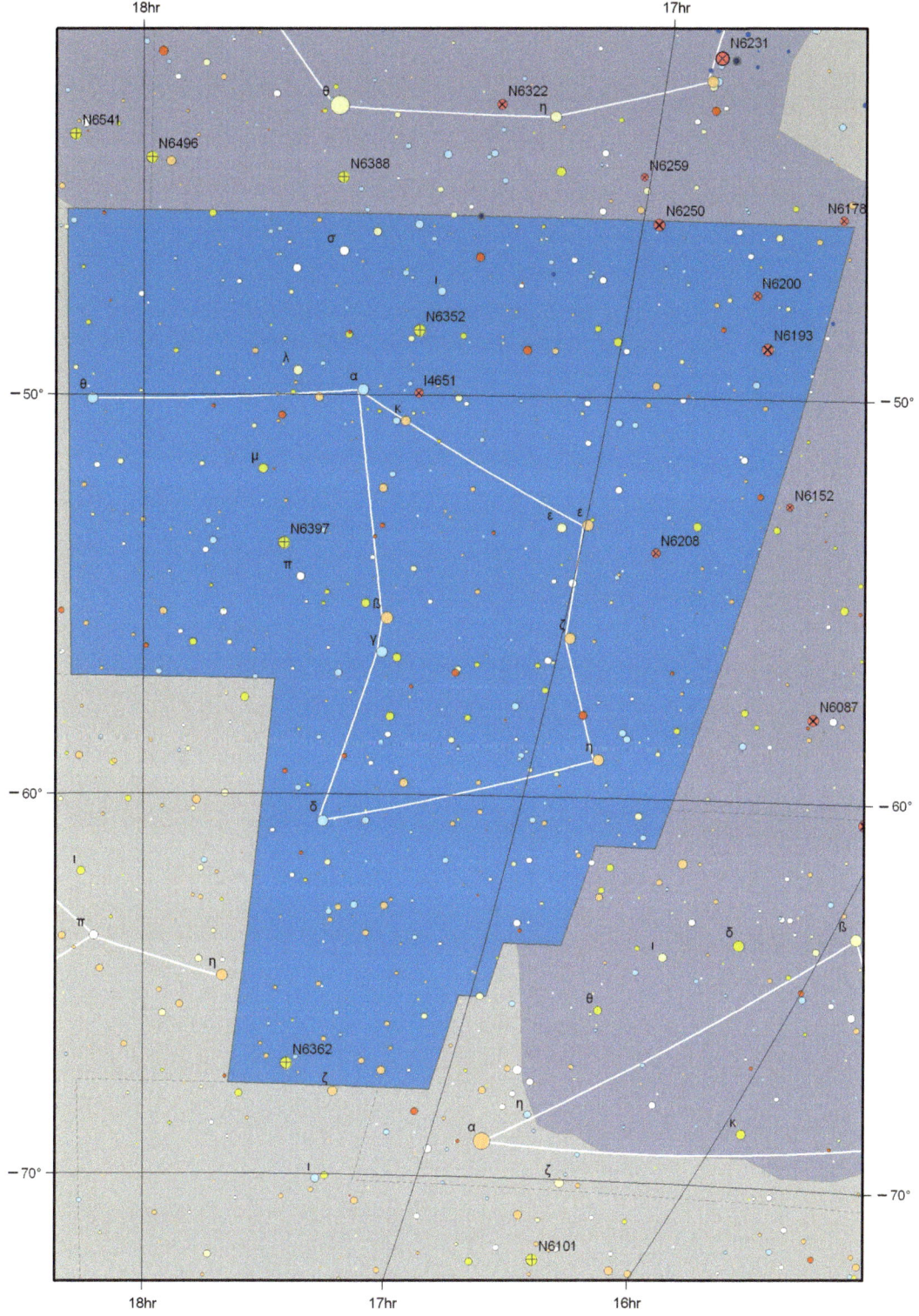

ARIES

A compact northern hemisphere constellation that is quite easy to spot because it lies in an area of sky – also encompassing Pisces, Triangulum and Cetus – within which bright stars are pretty much absent. Aries is best seen in July and may be viewed as a herald of the richer skies to come when Taurus and Orion begin to rise.

It is a zodiacal constellation and so can play host to any of the planets, the Moon or the Sun. When passing through, almost any planet will dominate the region, as the brightest 3 stars of Aries (alpha, beta and gamma) are of magnitudes 2.0, 2.6 and 3.9 respectively.

Aries used to contain the point at which the Sun in its passage across the sky passed from the southern celestial hemisphere into the northern hemisphere – the position of the Sun at the Vernal Equinox. However, because of the "wobble" of the Earth's axis caused by precession, the Sun now crosses the celestial equator further west, in Pisces. You will still find books that refer to the specific location as the First Point of Aries.

Aries exhibits two meteors showers during the year; unfortunately both appear during months when the sky around Aries is not dark, so we know them best from radar observations.

Historically

An old constellation representing a ram. The origins are a bit obscure, but it was inevitably appropriated by the Greeks to represent the ram from which the Golden Fleece sought by Jason came. The legend is long and detailed, with lots of incident – needless to say, the ram ended up dead.

Notable Stars

Double: Lambda Arietis 4.9/7.7 and 37″.

Double: Gamma Arietis 4.5/4.5 and 8″. Well matched A type white stars.

Deep Sky Objects

Aries lies well away from the Milky Way and is rather poorly served for deep sky objects. It does contain a few dim galaxies, but nothing worthy of note.

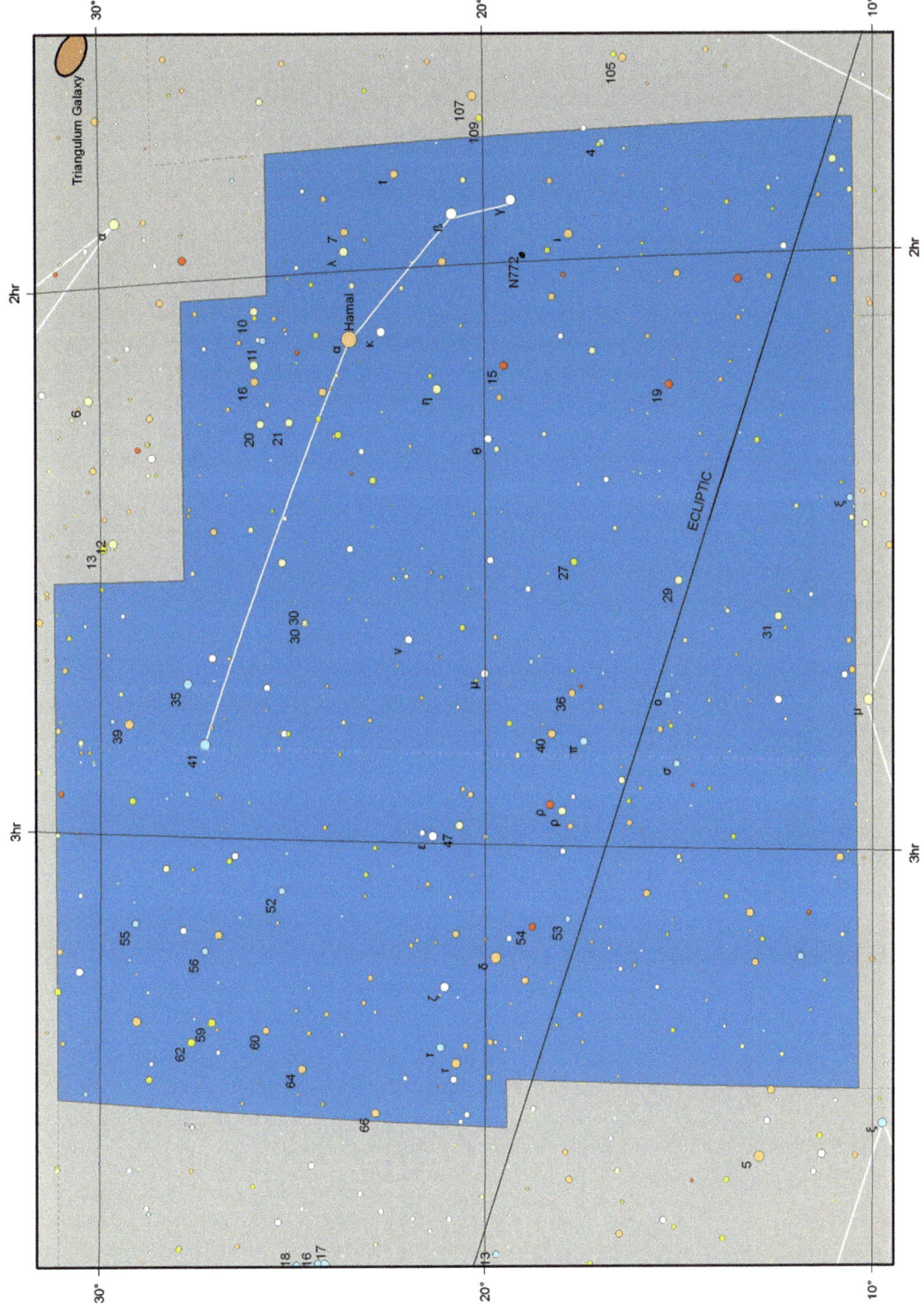

AURIGA

A northern hemisphere constellation dominated by the brilliant star Capella. Part of the Milky Way runs through the centre of Auriga and, as a consequence, it is home to a number of attractive and bright star clusters. It part-shares a star (gamma Aurigae) with Taurus, which was once known as beta Tauri, but has now been reassigned.

Auriga is a familiar feature of the skies during February and March.

Historically

The Charioteer is one of the 48 constellations identified by Ptolemy. But who it actually represents remains uncertain – Greek mythology is a fairly bloody affair providing lots of possibilities. Ignoring the chance that it's just a generic charioteer, it may derive from an elaborate tale about a king of Athens, Erichthonius, who was a very accomplished charioteer. But, oddly, the brightest star, Capella, means goat, and two of the stars nearby are known as the kids. But who would take a goat in a chariot? Maybe to pull a cart, perhaps? It seems the significance of the naming has been lost in the intervening centuries.

Notable Stars

Double: Theta Aurigae: 2.6/7.1 and 3.6″.

Double: Omega Aurigae 5.0/8.0 and 5.4″. Yellowish stars with another double star in the same field.

Variable: Eta Aurigae is a long period – 27 years – eclipsing binary where the nature of one of the eclipsing body is unclear – certainly a lot of dust orbiting the companion is involved. The most recent fade was in 2010.

Deep Sky Objects

M37: Of the four bright clusters in Auriga, *M37* is probably the most striking. It's a tightly packed collection of several hundred stars. The brightest is 8th magnitude and together they give the cluster an integrated magnitude of 6. It is resolved with even the smallest of instruments. Well worth returning to.

M36: A very attractive sight in any telescope. At around 25′ across and with far fewer stars than *M37*, it might sound unremarkable. It is not. Many of the cluster stars are quite bright, combining to give an integrated magnitude of 6. This makes the cluster bright and easy to find.

M38: This is probably the least attractive of the 3 Messier clusters. It is bigger and more sparse than its counterparts but, as a 7th magnitude cluster, is easily tracked down in binoculars. It is a nice find, containing several interesting chains of stars. It would probably be better appreciated if it was in less impressive company.

NGC1907: This cluster will appear in the same low-power field of view as *M38*.

NGC1931: This nebula/cluster will be a challenge for those of you using a 150mm aperture.

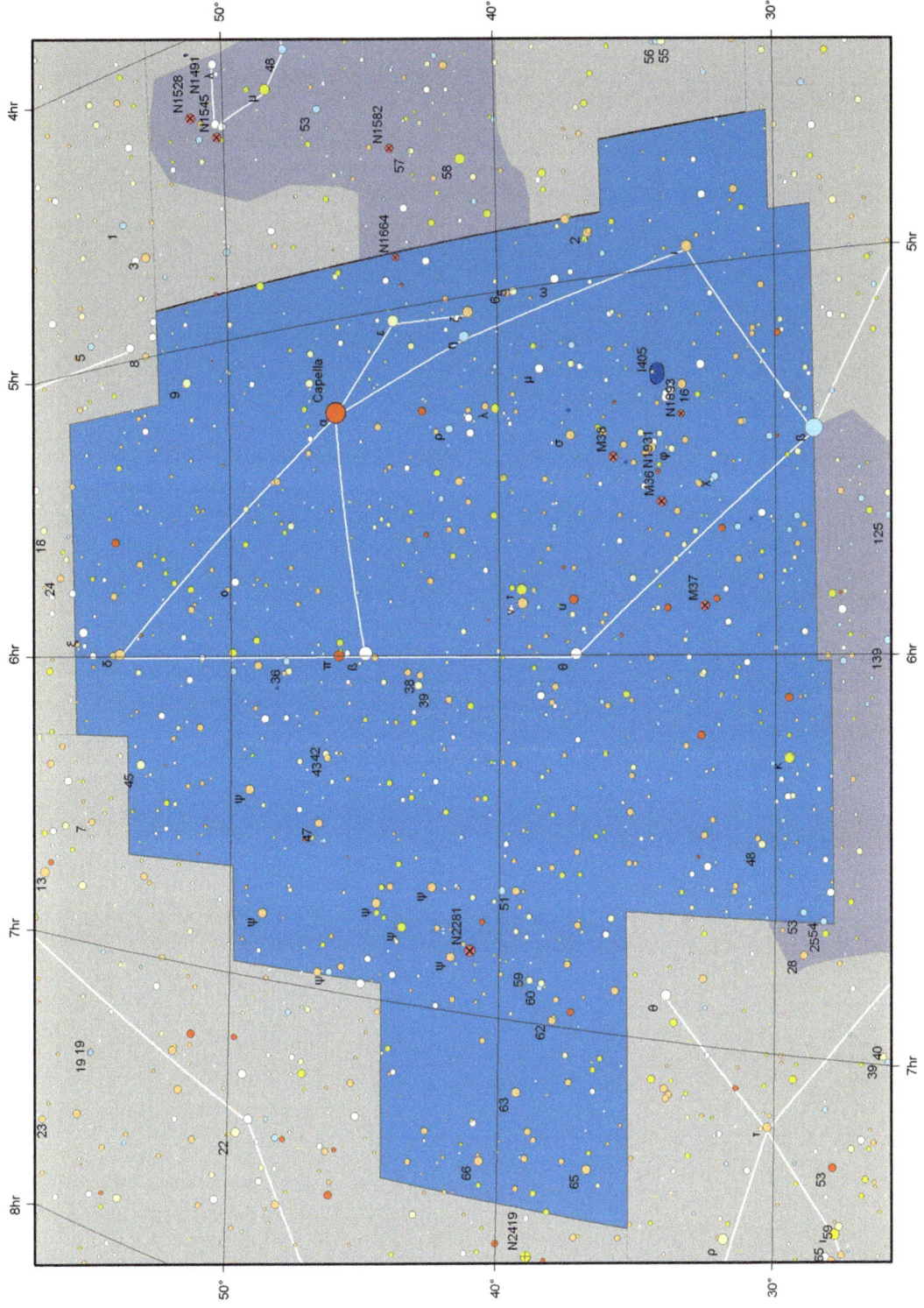

BOOTES

Bootes is a large and attractive Northern hemisphere constellation which is most easily traced out by starting with the orange giant star Arcturus.

The northern fringe of Bootes plays host to the Quadrantids meteor shower on January 4th. The duration of the shower maximum is usually short – only a few hours – but the activity rivals the Geminids. Other meteors may be seen on June 23rd – though this is not very reliable.

The constellation is a fine sight in the sky during the month of June.

Historically

An old constellation, the derivation of which is obscure, but likely to be a herdsman or ox driver wielding a stick or staff on his right side. It was recorded by Ptolemy in the second century and even gets a mention in Homer's The Odyssey, where Bootes is the driver of a wain or cart.

Just north of Bootes lies the defunct constellation of Quadrant Muralis. It depicted the quadrant instrument used by Lalande to measure star positions. Though it continued to appear in charts for a century or so, it was not adopted formally in 1930 when the constellation boundaries were finalized. These days the name only persists in the name Quadrantids, mentioned previously.

Notable Stars

Double: Kappa Bootis 4.6/6.6 and 13.4″. White and blue.

Double: Iota Bootis 4.9/7.5 and 38″. Yellow and blue stars.

Double: Delta Bootis 3.5/8.7 and 105″. Yellow stars.

Double: Mu Bootis 4.3/7.0 and 108″. The fainter component is itself double consisting of two orange 7th magnitude stars just 2.2″ apart.

Double: Epsilon Bootis 2.5/5 and 3″. Pale yellow and bluish.

Double: Xi Bootis 5.5/7.8 and 7″. A mere 22 light years away. It's one of the nearer sun-like stars, but a lot younger. Yellow and orange.

Deep Sky Objects

There are no notable deep sky objects in Bootes.

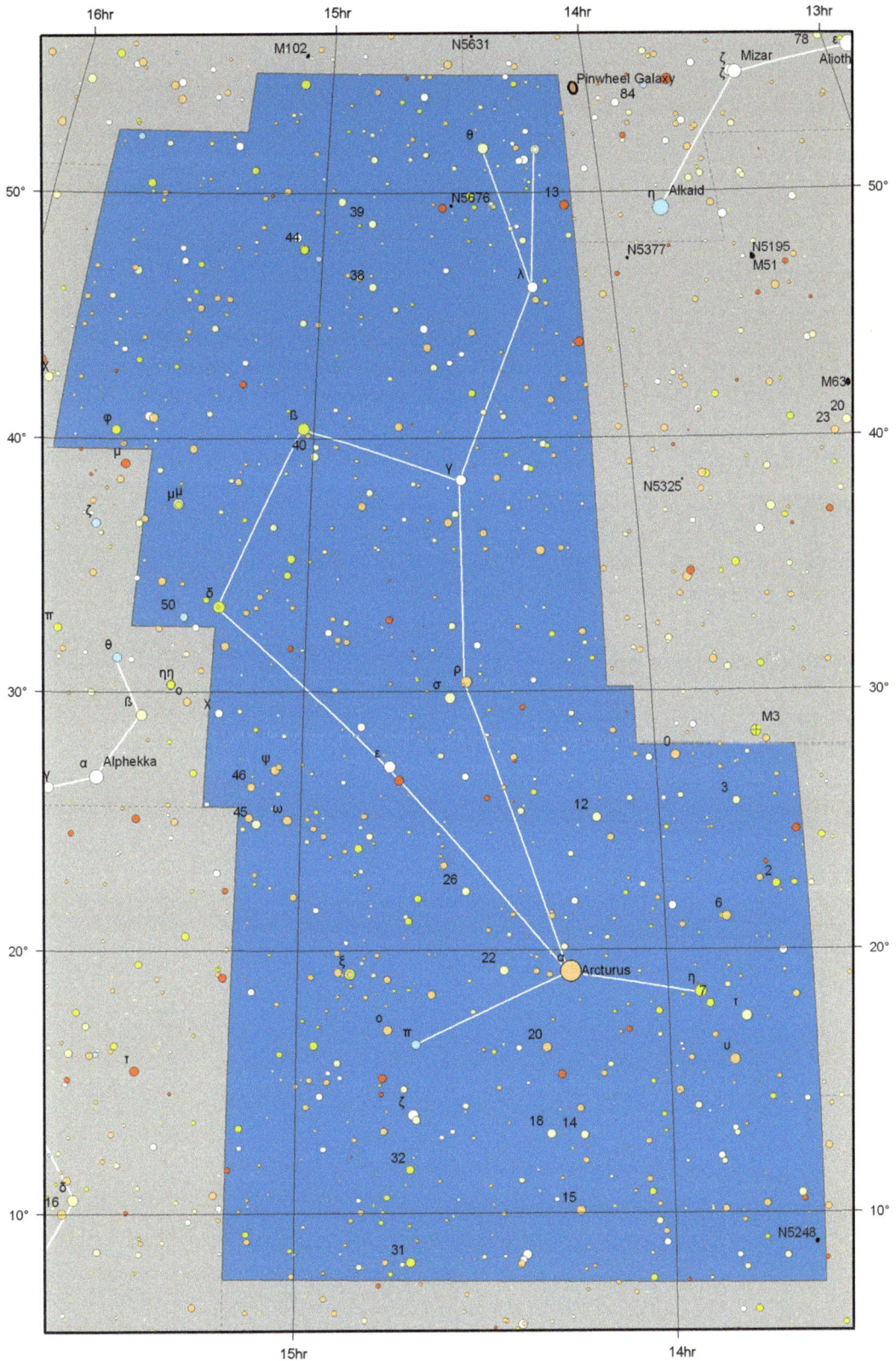

CAELUM

A small and not very impressive southern hemisphere constellation to be found just to the west of Columba. None of its stars are very bright, being 4th magnitude or fainter. There's no obvious form to the constellation. which is scarcely more than a couple of isolated stars and one of the smallest in the sky.

Best placed for observation during January and the early part of the year.

Historically

This is another constellation created by Nicolas Louis de Lacaille: one of 15, in fact. It represents a type of engraving tool and a sculptors' chisel, bound in ribbon. It appeared as les Burins on his original charts, before being changed to Caelum Sculptorium. De Lacaille was clearly fascinated by the technical apparatus of the time, as several of his constellations refer to instrumentation of one type or another.

Notable Stars

Double: Alpha Caeli 4.5/12.4 and 6.6″. A challenging double where the dimmer star is a red dwarf. The brighter star is F type and may be slightly variable.

Double: Gamma Caeli 4.5/6.3 separated by 13 arc minutes. The fainter of the two is itself double, with an 8th magnitude companion lying 3″ away.

Deep Sky Objects

There are no notable deep sky objects in Caelum.

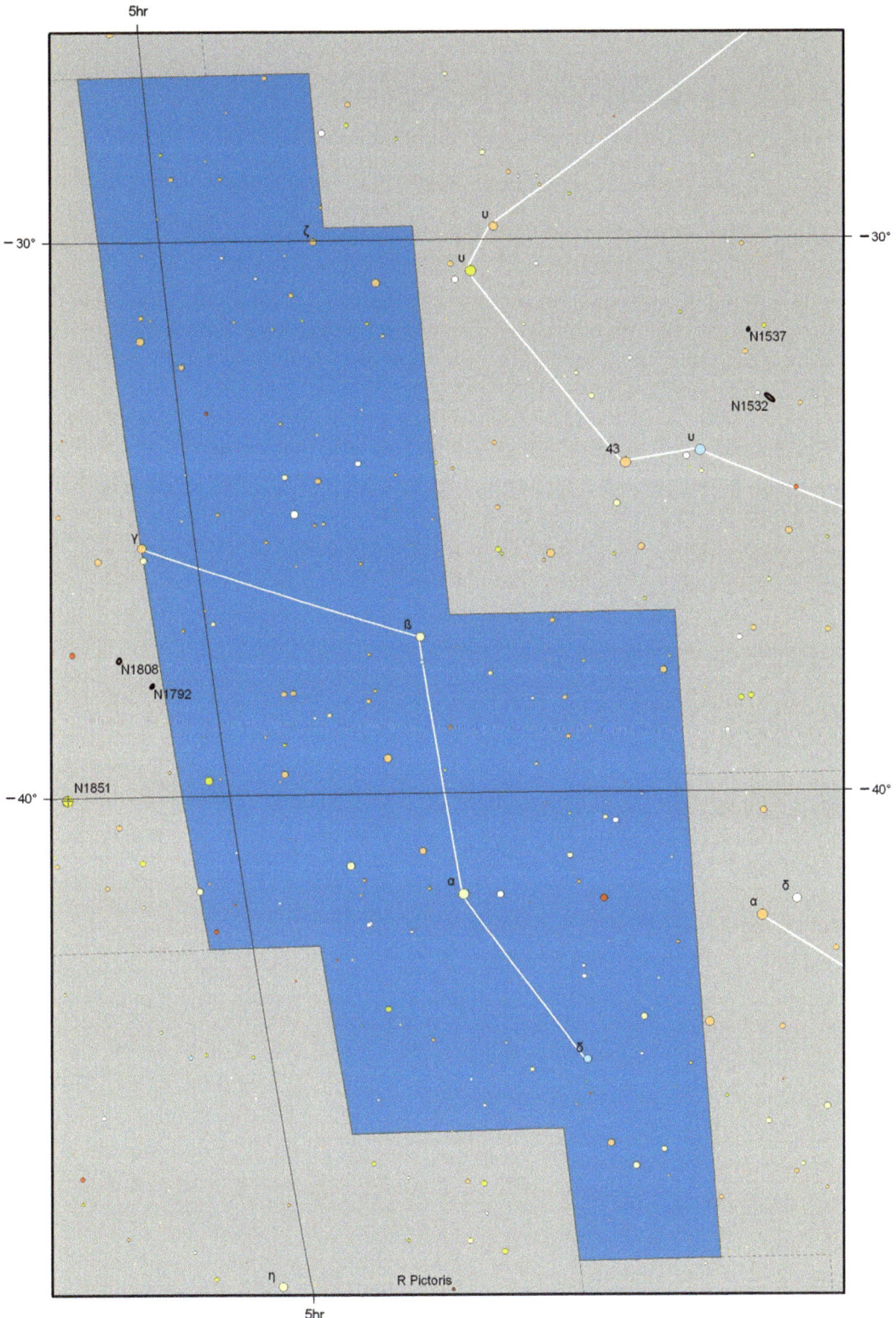

5hr

υ

−30°

ζ

υ

N1537

N1532

−30°

43

υ

γ

ß

N1808

N1792

N1851

−40°

−40°

α

δ

α

δ

η

R Pictoris

5hr

CAMELOPARDALIS

A large constellation, covering a lot of sky, quite close to the North Celestial Pole. It sprawls between Capella in Auriga and Polaris. Its brightest stars are 4th magnitude and the outline is not very memorable, but, skirting the edge of the Milky Way, it does contain a few interesting targets for observation.

It is best seen during the month of February.

Historically

An invention of the seventeenth century astronomer Petrus Plancius, Camelopardalis depicts a giraffe and not, as you might expect, a camel. Clearly, the creature really impressed Plancius, as no legend is associated with a giraffe – though there might have been if Zeus had known about stepladders. Lots more of him later.

Nearby was the, now obsolete, constellation of Tarandus, which represented a reindeer. Given the lack of bright stars in the area, Camelopardalis probably survived only because of its size – the 18th largest constellation.

Notable Stars

Distant: Alpha Camelopardalis is a super-luminous blue O9 type supergiant very remote from us, lying some 6,000 light years away.

Double: Beta Camelopardalis 4.0/85″. Yellow and blue stars. Actually the brightest star in the constellation, despite being 1,000 light years distant. The fainter companion is itself a double star, with the additional 11th magnitude star some 14″ away.

Double: 7th magnitude triple star in open cluster NGC1502

Deep Sky Objects

NGC1502: An open cluster of 20 or so stars roughly 8 arc minutes across.

Kemble's Cascade: An eye-catching run of stars with *NGC1502* at one end. Look for a number of unrelated stars aligned north west from *NGC1502*. A relatively recent discovery, it was first observed with binoculars just 32 years ago.

NGC2403: A 9th magnitude galaxy roughly 12×20 arc minutes in diameter. A 150mm aperture will capture it, but details require much larger apertures.

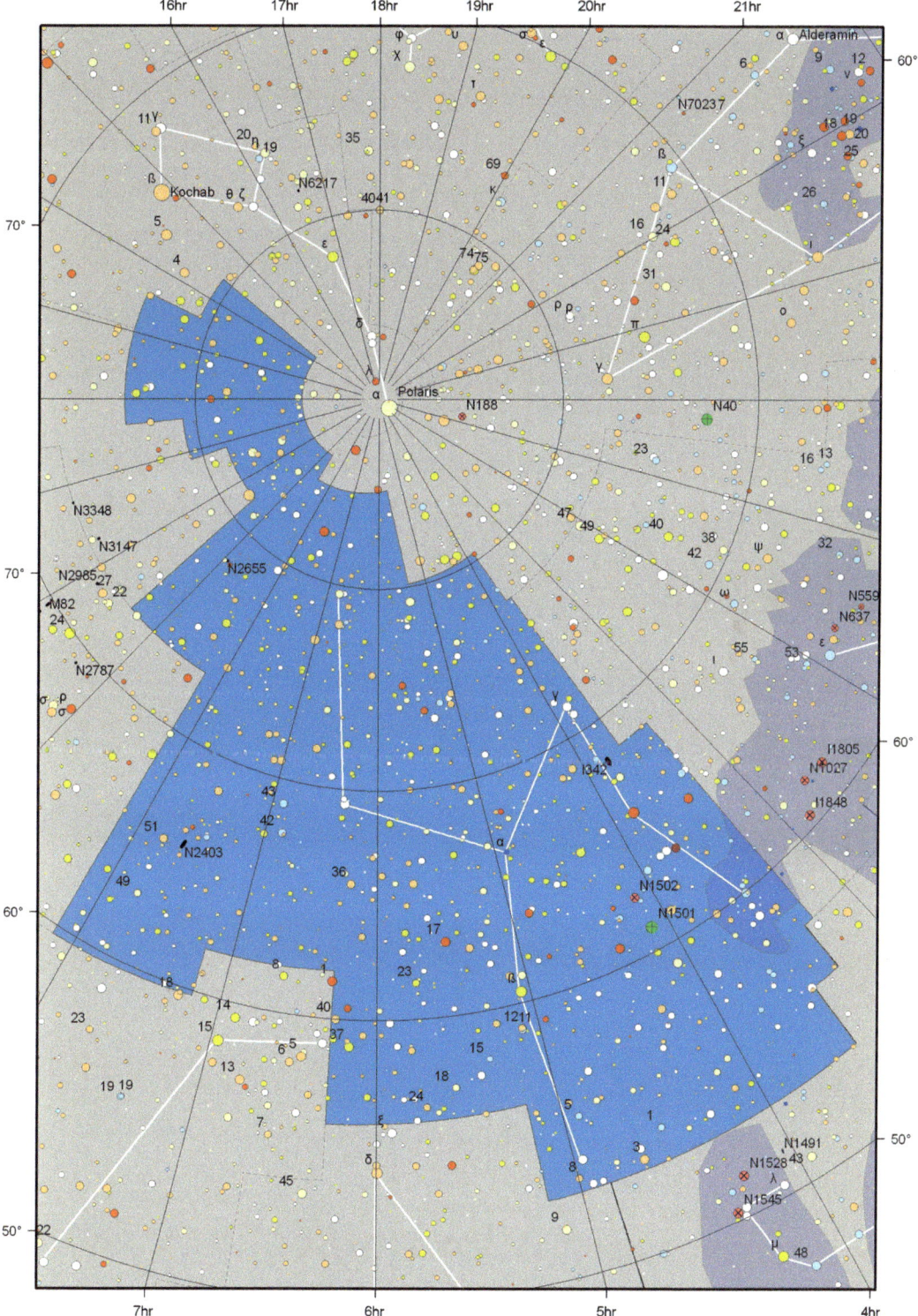

Fig. 3 Here is Kembles Cascade, a line of stars visible in binoculars and small telescopes that lie in Camelopardalis (Image by Greg Parker)

CANCER

A not very bright northern hemisphere constellation made up of 3rd, 4th and 5th stars. It's not close to the Milky Way but, nonetheless, contains two splendid clusters, one of them visible to the unaided eye. The ecliptic passes through Cancer, nearly crossing one of the clusters.

Historically

The best theory is that this zodiacal constellation derives from the 12 labors of Hercules. Apparently, a rather reckless crab tried to intervene in the fight between Hercules and the Hydra. Unsurprisingly, the crab got the worst of the encounter and was either kicked into the sky or placed there by the grateful Hera, who despised Hercules. It's one of the least impressive of the 12 classical zodiacal constellations.

Notable Stars

Double: Zeta 5.5/6.0 and 6″. A pair of yellowish stars not far from M44.

Double: Iota Cancri 4.4/6.5 and 31″. Colors bluish and orange, providing a pretty contrast.

Deep Sky Objects

M44: The Praesepe or Beehive cluster nestling at the centre of Cancer. It is 3rd magnitude and thus visible with the naked eye. The brightest star is magnitude 6 and the cluster is best seen in a low power view with something like a 114mm f5 reflector (or good binoculars) as it appears wider than the full moon. It has been known since antiquity – Ptolemy described it as a "nebulous mass" – so it hardly counts as a discovery by Messier. It lies roughly 600 light years away from us.

M67: Another bright cluster. The light from the stars of M67 combine to provide a cluster magnitude of 6. It is quite an old star cluster – four billion years – and hence has been swept clean of the dust and debris associated with younger clusters like the Pleiades. Binoculars will resolve a couple of its brightest members. All its heavier and brighter stars have burned up their hydrogen fuel and left the main sequence.

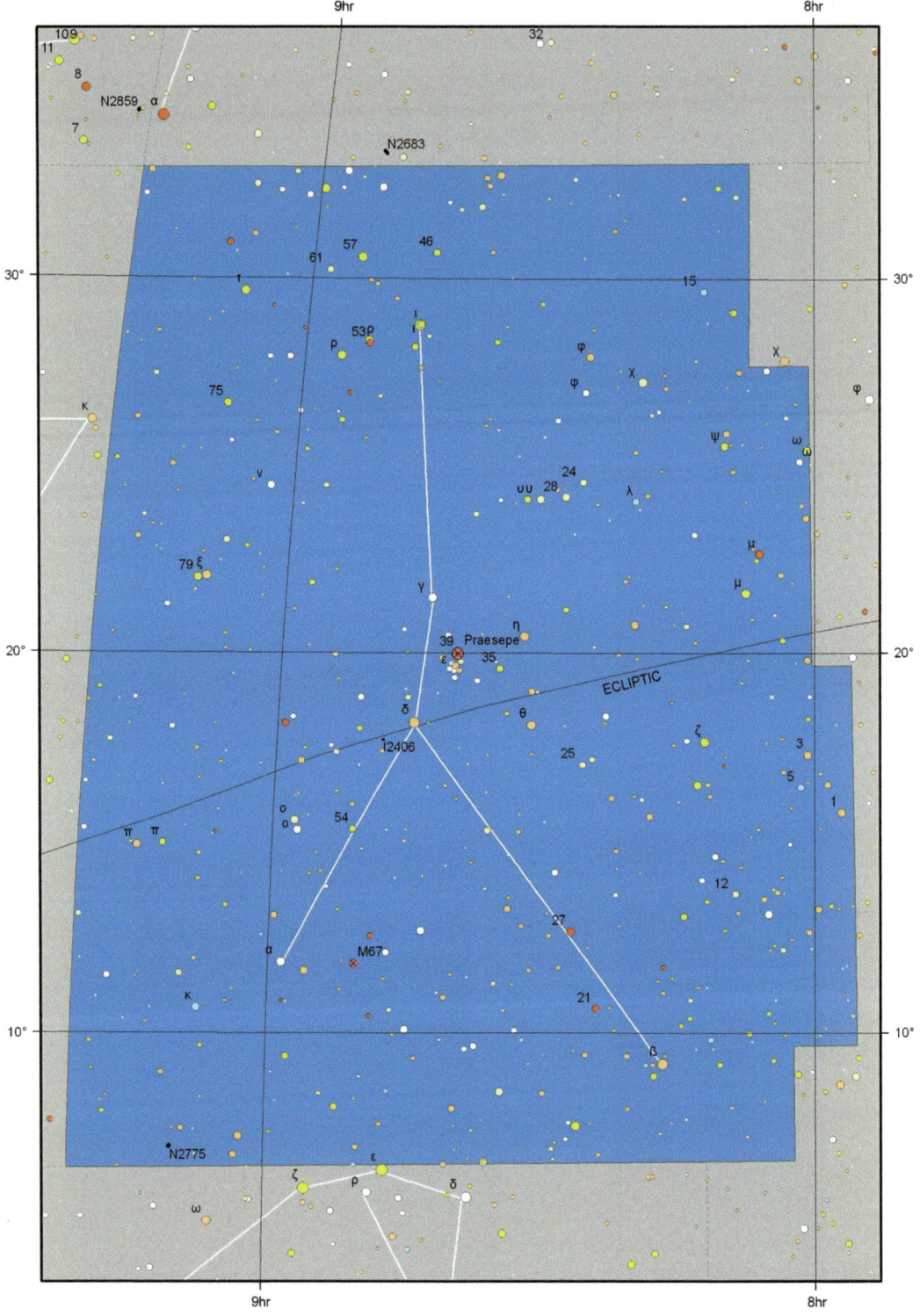

Fig. 4 M67, an attractive open cluster in Cancer often overlooked because that small constellation also contains M44 (Image by David Ratledge)

CANES VENATICI

A northern hemisphere constellation that can be found just south of the eastern end of The Plough in Ursa Major. Although there are few bright stars within its form, one of them is Cor Caroli, a bluish 2nd magnitude star named by Charles Scarborough, one of the first Fellows of the Royal Society. He named it after the executed King Charles I and, subsequently, became Sir Charles Scarborough, knighted by King Charles II a few years after the restoration of the British monarchy.

Canes Venatici is best seen during the month of May.

Historically

The constellation was drawn by Johannes Hevelius in 1687 and represents two hunting dogs controlled by Bootes. The two stars alpha and beta Canes Ventaticorum represent the southernmost pair. Dogs had appeared on earlier star charts, but the version created by Hevelius made more use of the stars immediately below Ursa Major and is still recognized.

Notable Stars

Double: Alpha Canes Venaticorum 3.0/5.5 and 20″. Pale yellow stars.

Sunlike star: The star Beta Canes Venaticorum is a sun-like star – a G type dwarf on the main sequence. Easily visible sun-like stars are rare, with only a few dozen scattered across the sky. This is probably the easiest to spot in the northern hemisphere of the sky.

Deep Sky Objects

M3: A fine globular cluster on the southern edge of the constellation. Imaged, its half-a-million stars appear as a sphere 18 arc minutes across. It can be partially resolved in a 100–150mm scope, and is a splendid sight in a large instrument. *M3* can be seen in binoculars as a small, slightly blurred star-like point, of magnitude 6.

M51: A lovely deep-sky target – The Whirlpool Galaxy. This 8th magnitude face-on spiral and its peculiar companion *NGC5195* are an ideal target for any observer. They are detectable in 10×50 binoculars or finder scopes at a pinch. A 150mm shows that the galaxy *M51* appears a little granular, while a 250mm will – under very good conditions – show hints of the spiral arms and the star forming regions they contain. Earl Rosse used a 72-in. metal mirror to detect the arms originally, but modern optics make the structure discernible using much smaller instruments

M63: A large and bright (8th magnitude) inclined spiral galaxy. Although *M63* can be glimpsed with a 60mm telescope, none of its fine spiral structure can be seen visually with anything but the largest scopes. An 8th magnitude star lies nearby.

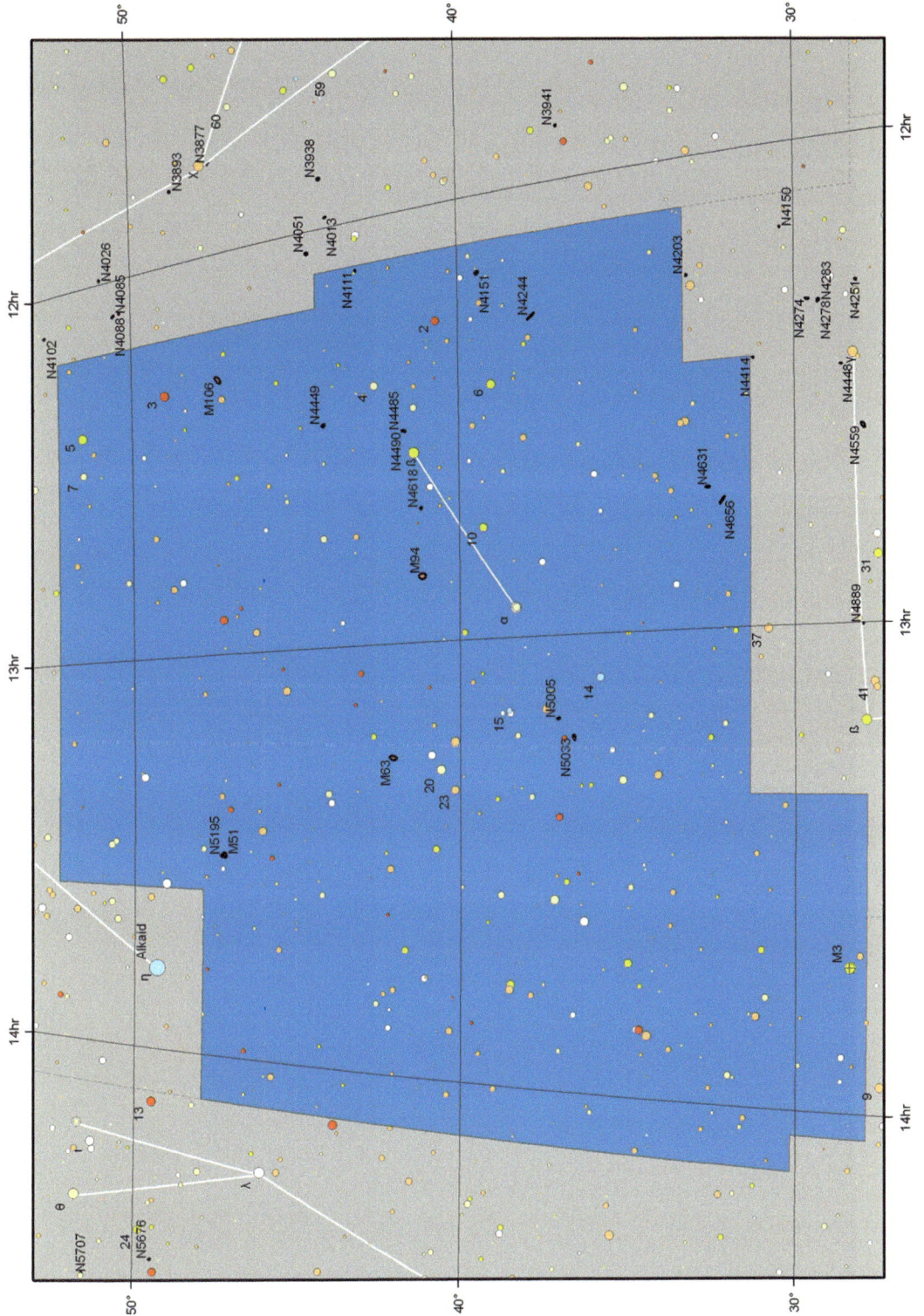

Fig. 5 The Whirlpool Galaxy, a spectacular face-on spiral galaxy found, together with its neighboring galaxy, in Canes Venatici (Image by David Ratledge)

Fig. 6 M63 in Canes Venatici is located close to an 8th magnitude star making locating it particularly easy (Image by Grant Privett)

CANIS MAJOR

An impressive southern hemisphere constellation containing the brightest star in the sky, Sirius. It can be located just south east of Orion and the two constellations provide an superb pairing that is best seen from south of 40N latitude, as, in northern latitude, It's never particularly high above the horizon. Canis Major is just west of the Milky Way, but the fields are attractive for low power sweeping.

It is highest in the sky on late evenings in February.

Historically

Like Canes Venatici, Canis Major represents a dog, but in this case made up of far more impressive stars, and trailing Orion rather than preceding Bootes. It is dominated by Sirius, the Dog Star. When Sirius was seen to rise in the predawn, the hottest part of the northern summer was on hand and, in Egypt, this information was used to help time the irrigation of the fields.

The dog represented may be Laelaps, a hugely successful hunting dog, but there are other contenders. Some view it as hunting the hare, Lepus, that can be seen below the feet of Orion.

Notable Stars

Double star: Alpha Canis Majoris (Sirius) –1.4/8 and 7″. The separation will widen until 2022 when they will be 12″ apart.

Double: Epsilon Canis Majoris 1.5/7.4 and 7.5″. The brighter star is very blue and the brightness difference makes this difficult to split.

Double: Tau Canis Majoris 4.5/10 and 8″ lies within the open cluster NGC2362.

Nearby star: Canis Majoris (Sirius) is the brightest in our sky but isn't intrinsically bright. It appears so bright because it and its companion are a mere 8.5 light years removed from us.

Deep Sky Objects

M41: A beautiful open cluster. Despite its low altitude as viewed from the UK, M41 is still an impressive sight in almost any instrument 4° south of Sirius. It appears as a cluster of some 25 stars, 10 of which are brighter than magnitude 8.5. It covers about 30 arc minutes of sky. Even modest instruments will be able to discern the string-like patterns of stars that make the cluster so attractive. Look out for a yellowish-red star near the centre and a brighter star on the southern edge. A visit to M41 is well worth the effort.

NGC2362: A smallish cluster (diameter 5′) yielding 10 or so stars to smaller instruments and many more to larger apertures. Many of the members appear blue in color but at a declination of –25° it may be too near to the southern horizon for many observers at northerly latitudes.

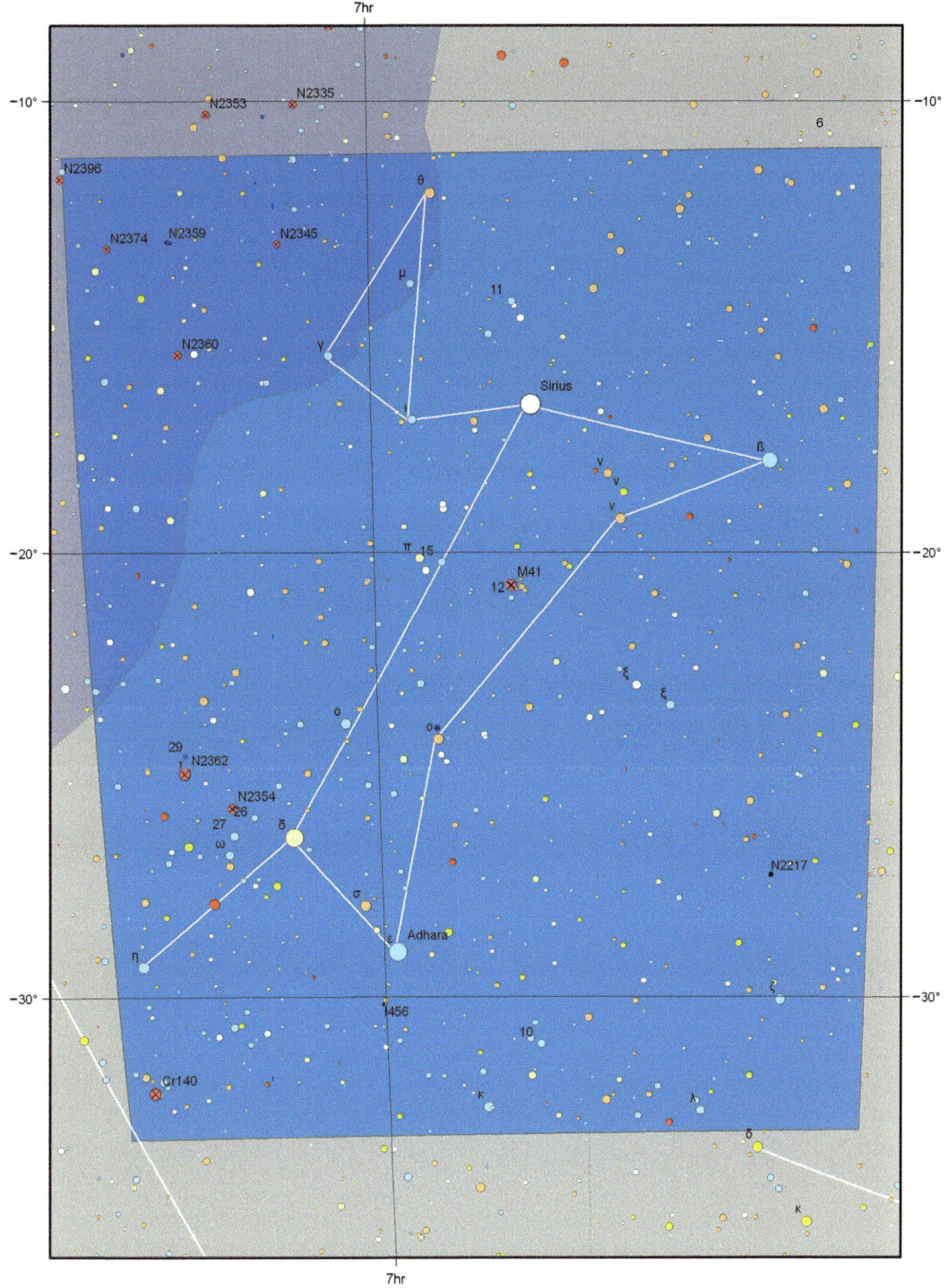

7hr

−10°   −10°

N2353   N2335

6

N2396

θ

N2374   N2359   N2345

μ

11

γ

N2360

Sirius

ß

ν

ν

ν

−20°   −20°

π 15

M41

12

ξ

ξ

o

o

29 N2362

N2354

26

27

δ

ω

N2217

σ

η

ζ

Adhara

−30°   −30°

1456

10

Cr140

κ

λ

δ

κ

7hr

CANIS MINOR

A constellation very similar to Canes Venatici, but with an even brighter leading star. Alpha Canis Minoris is Procyon, a yellowish 0th magnitude star. The origins of the names suggest that its rising was a precursor to a more important event, the rising of Sirius. The constellation is on the edge of the Milky Way, but despite this it contains little else of note.

Procyon and the surrounding constellations are to be found riding highest in the sky during March.

Historically

Unsurprisingly, another dog. This small (minor) dog is certainly less impressive than Canis Major but boasts a bright star in Procyon. One legend has it that the hound can be identified with a dog belonging to an Athenian called Icarius who, legend has it, was the first man taught to make wine. He shared a few bottles with some shepherds and then, unused to being intoxicated, they murdered him fearing they had been poisoned. The dog alerted the family of Icarius, who promptly killed themselves. Jolly.

Notable Stars

Nearby star: Alpha Canis Minoris (Procyon) is one of the nearest stars to us at 11.4 light years away. Procyon is closely orbited by a white dwarf star.

No double stars

Deep Sky Objects

There are no deep sky objects of note within Canis Minor.

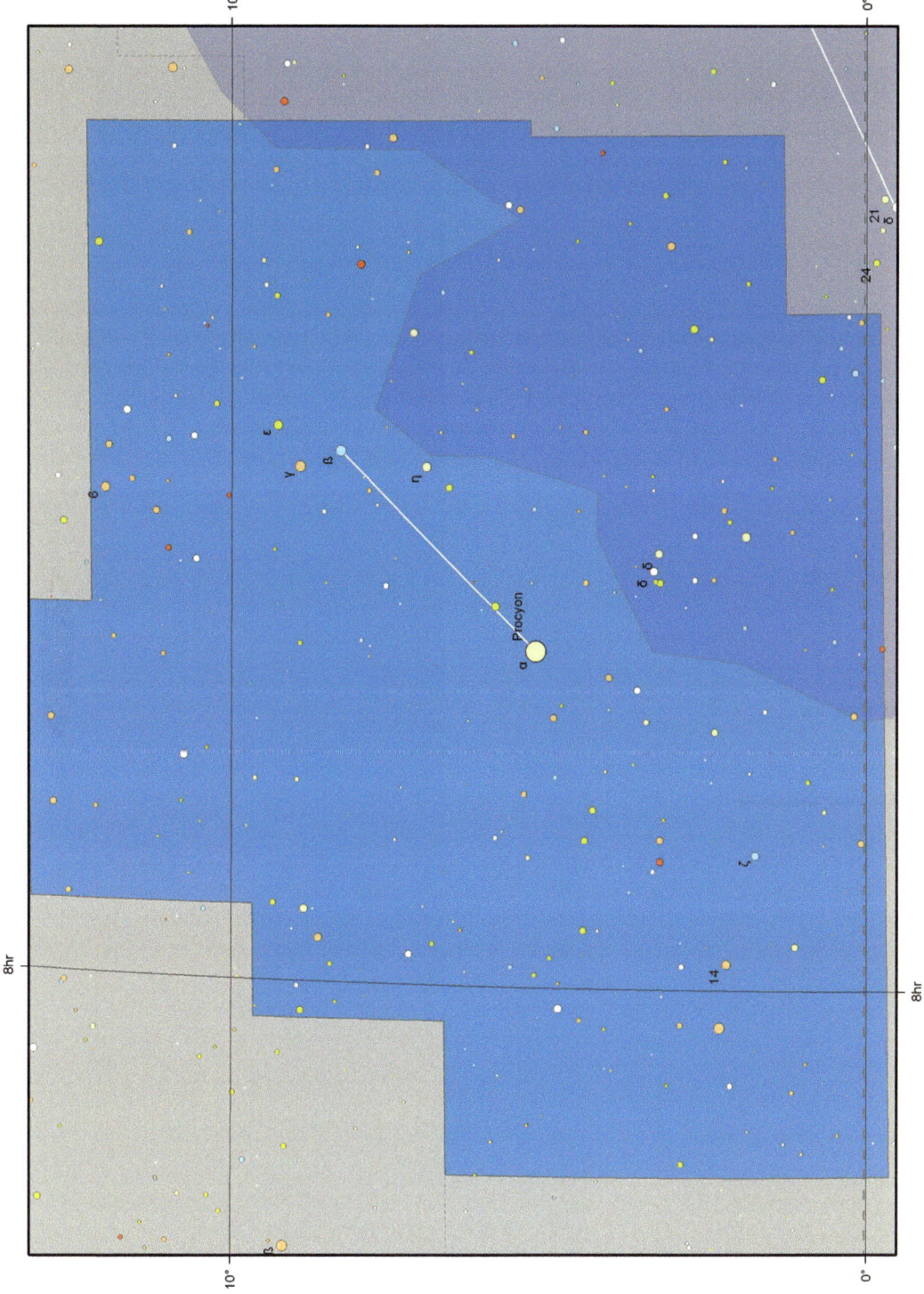

CAPRICORNUS

A constellation of the southern hemisphere. As an ancient zodiacal constellation it inevitably has the ecliptic running through it and will play host to the Moon every month. Its brightest star is delta Capricorni which is 2nd magnitude. Its shape is quite distinctive.

It can be located between Altair and Pisces Austrinus.

Historically

Another constellation of the classical zodiac. It's pretty ancient and represents a creature with the head of a goat and the tail of a fish – sometimes referred to as a sea goat. It may have had its debut as long ago as the Sumerians. Some Greeks saw it as Pan, who was a bit of a wild God and not very nice – and that's against some pretty stiff competition. He wandered the fields, forests and hills generally having a good, if rather rowdy and lawless, time with a bunch of like-minded nymphs.

Notable Stars

Double: Alpha Capriconi 3.6/4.3 and 6.3 arc minutes. A line of sight double star. One of these yellow stars is nearly 600 light years further away from us than the other. The name alone, *Giedi Prime*, is enough to guarantee *Dune* readers give it a look.

Double: Beta Capricorni 3.0/6.1 and 3.5 arc minutes. The dimmer star has its own companion some 6 arc minutes away.

Deep Sky Objects

M30: This globular cluster in the south eastern part of the constellation is the only notable deep sky object. Through smaller instruments M30 appears about 7 arc minutes wide and it is of 8th magnitude. It is visible as a slightly fuzzy star in 10×50 binoculars from locations where its −23 declination puts it high in the sky. Unusually, it orbits our galaxy in the opposite direction to most clusters, suggesting we may have acquired it from a satellite galaxy at some point.

Having observed M30, why not sweep north to the globular cluster M15 in Pegasus?

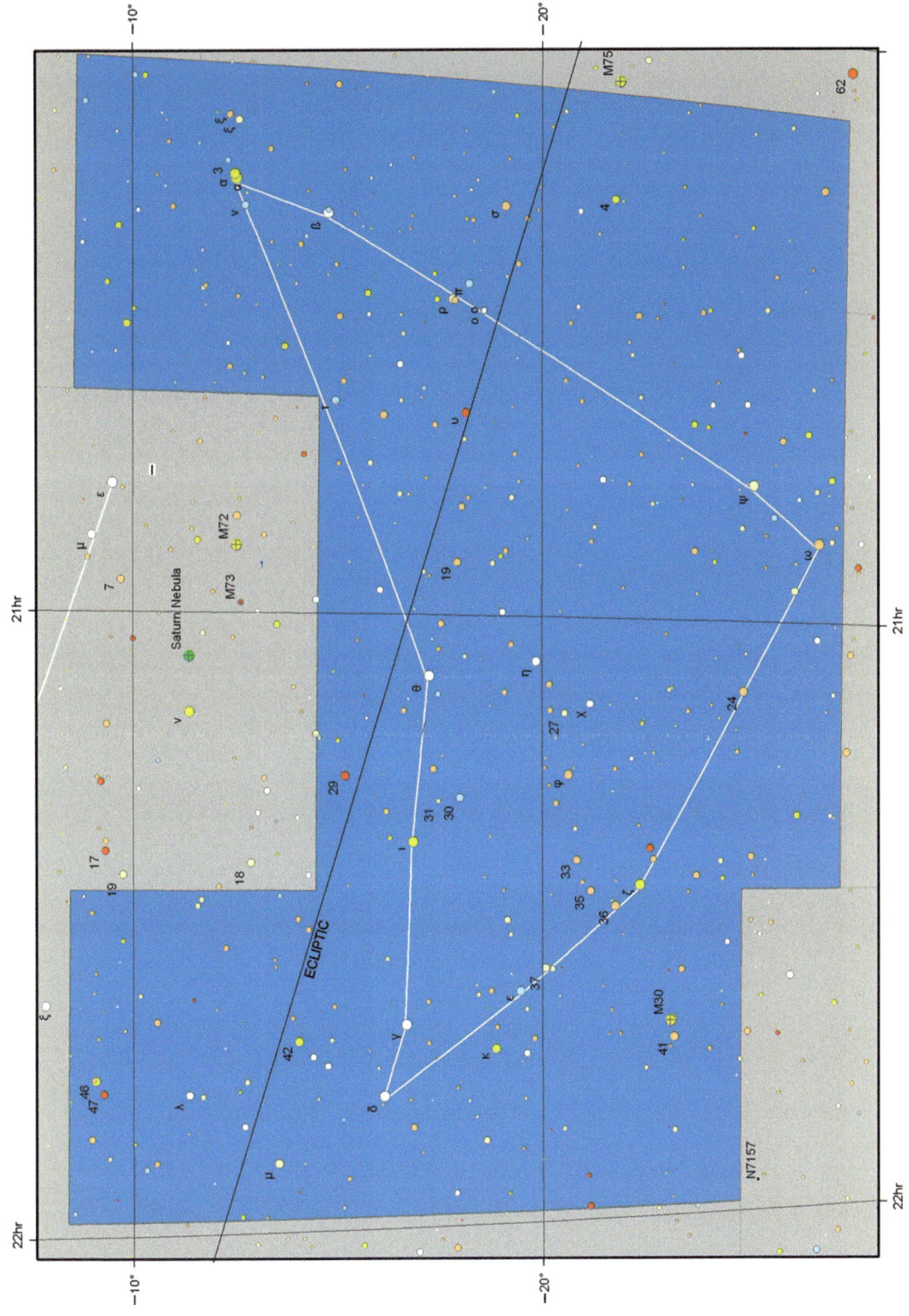

CARINA

A beautiful southern hemisphere constellation in one of the most lovely regions of sky. Its position within a rich portion of the Milky Way provides several bright clusters and also the enigmatic nebula which surrounds the increasingly unstable star, eta Carinae. The constellation is dominated by the brilliant Canopus. The eastern end of the constellation adjoining Crux is full of interest and is reason alone to believe Australia is the "lucky country".

A constellation best enjoyed around March.

Historically

The keel of the ship. In this case the keel (or hull) of the ship sailed by Jason which was originally depicted by the enormous, and now obsolete, Argo Navis constellation. It's one of three sections, the others being Puppis, the poop deck (stern) and Vela, the sails. Pyxis, the nearby ships-compass constellation, is not related, being a far later invention. Argo was broken up by Nicolas Louis de Lacaille in 1756.

Notable Stars

Double: Upsilon Carinae 3.0/6.0 and 4.8″. Blue tinted stars. There is a pair of 8th and 9th magnitude stars separated by 12″ in the same field of view making the effort doubly worthwhile.

Variable: Eta Carinae has spent much of the last 130 years at about 8th magnitude, but occasionally flares up to 1st magnitude for a few years. It can be thought of as a supermassive star preparing to go supernova. It's a ticking bomb and will probably explode within the next few millennia. The star is believed to be a close double star and is four million times brighter than our Sun. Keep watching!

Deep Sky Objects

NGC2808: A 7th magnitude globular cluster some 4 arc minutes across. It's one of the most massive globular clusters. Under favorable conditions a 200mm aperture will resolves some of its one million stars. It lies 31,000 light years away.

NGC3372: The wonderful eta Carina nebula is one of the highlights of the southern sky. Eta Carina, at its core, is shrugging off shells of gas and dust that are visible to the naked eye on a dark night and which make a lovely sight at the eyepiece. It's bigger and more attractive than the Orion nebula and contains regions of dark nebulosity that make drawing it challenging.

IC2602: A fine naked eye cluster sometimes known as the Southern Pleiades. It is obvious to the naked eye, with its 50 stars strewn over nearly the degree of sky surrounding theta Carinae. A low power is essential and the view in binoculars well worth seeing. The whole area around here is wonderful.

NGC2516: An open cluster clearly visible to the naked eye. This rich cluster covers about half a degree of sky. Its aggregated magnitude is 3.8.

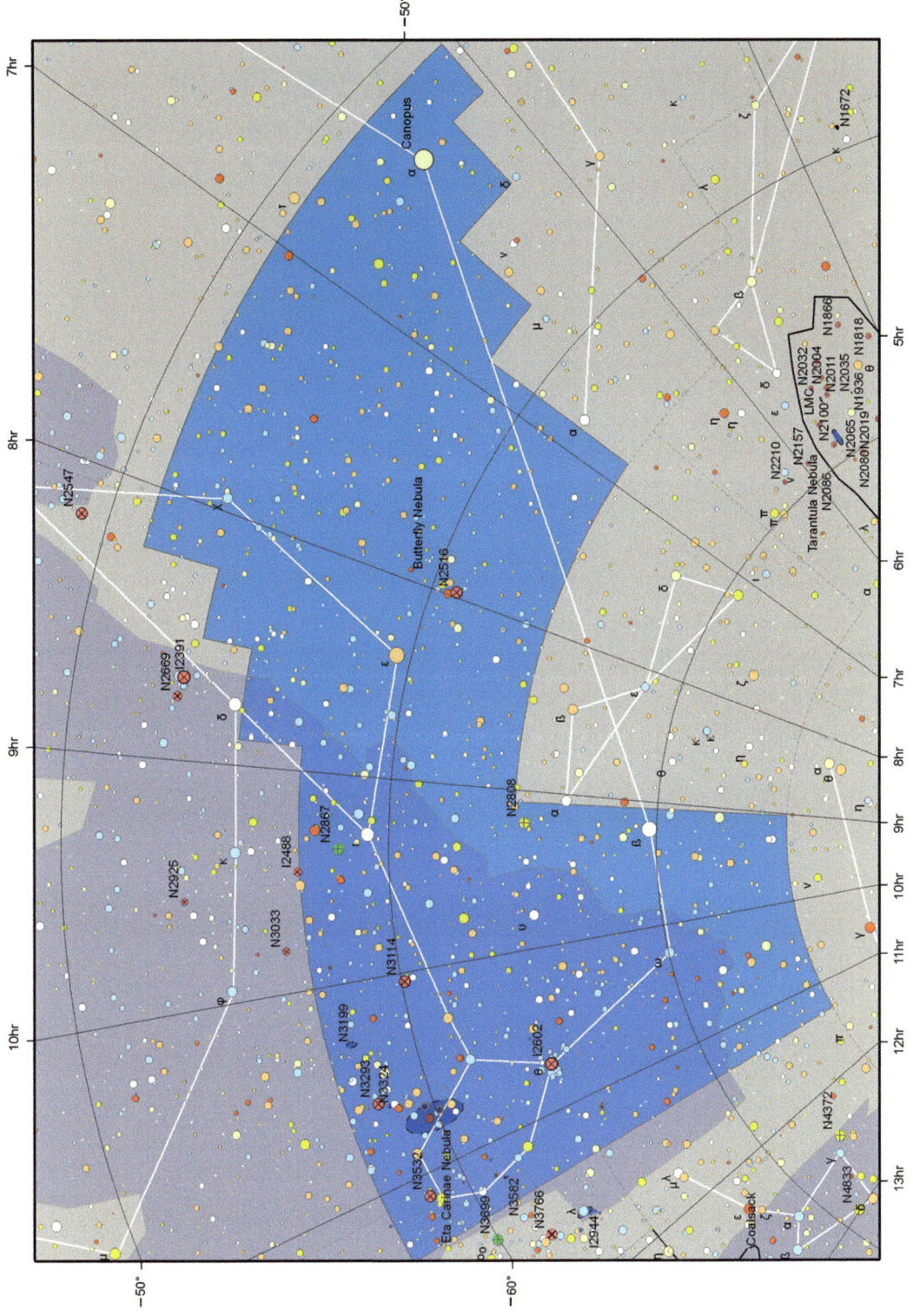

Fig. 7 The wonderfully detailed and attractive eta Carinae nebula marks a star poised on the edge of becoming a supernova (Image by Chris Picking)

Fig. 8 The Southern Pleiades cluster in the constellation of Carina. A gem of the southern sky (Image by Chris Picking)

CASSIOPEIA

An attractive and readily spotted northern hemisphere constellation lying on the other side of Polaris from the pointer stars of Ursa Major. Its "W" or "M" shape – depending upon the time of year – is instantly memorable. In a dark sky, the constellation is seen against the Milky Way and thus is even more impressive. Sweeping across the region with a small telescope is very worthwhile, as there are numerous lovely fields of view to get lost in.

A constellation seen at its best in November.

Historically

Cassiopeia was the wife of King Cepheus of Ethiopia. Apparently the lady was a bit of an airhead and proclaimed that she was beautiful – which was fine – and even more beautiful than the sea nymphs, known as the Nereids – which was very much not fine. Apparently, being vain is permitted for gods but vanity in mortals is not, so 50 Nereids went looking for her, and a sea monster was sent to ravage the kingdom – the coast at least. Cassiopeia and Cepheus, caring parents evidently, had their daughter Andromeda strapped to a rock as an appeasing sacrifice. Fortunately, Andromeda was rescued in the nick of time by Perseus, thereby thwarting the plans of the sea God.

Notable Stars

Double: Alpha Cassiopeiae 2.5/8.9 and 64″. Red and blue components.

Double: Eta Cassiopeiae 3.4/7.5 and 11″. Yellow and red stars.

Double: Iota Cassiopeiae 4/7/8 2/7 241″/114″. Triple star.

Double: Sigma Cassiopeiae 5.2/7.1 and 3″ Blue and yellow stars. Quite tough to separate in a small 'scope.

Variable: Gamma Cassiopeiae is a rapidly spinning eruptive variable and blue star. It can be brighter than alpha and beta Cassiopeiae when it is shedding material into a shell of gas. It is exhausting its hydrogen, and transitioning toward becoming a giant star. It varies between 1.6 and 3.4.

Deep Sky Objects

M103: A bright and compact cluster of stars lying 8,000 light years away. Contains around 40 stars. Easy to locate near delta Cassiopeiae.

NGC663: A reasonably rich 7th magnitude open cluster, boasting 100+ members that, in amateur telescopes, sprawl over 8′ in a rather clumpy manner. This young cluster is well seen in binoculars.

Stock2: A large and striking, but sparse, cluster standing out from the field stars. A degree wide, it is best viewed with binoculars. It's a very pretty sight.

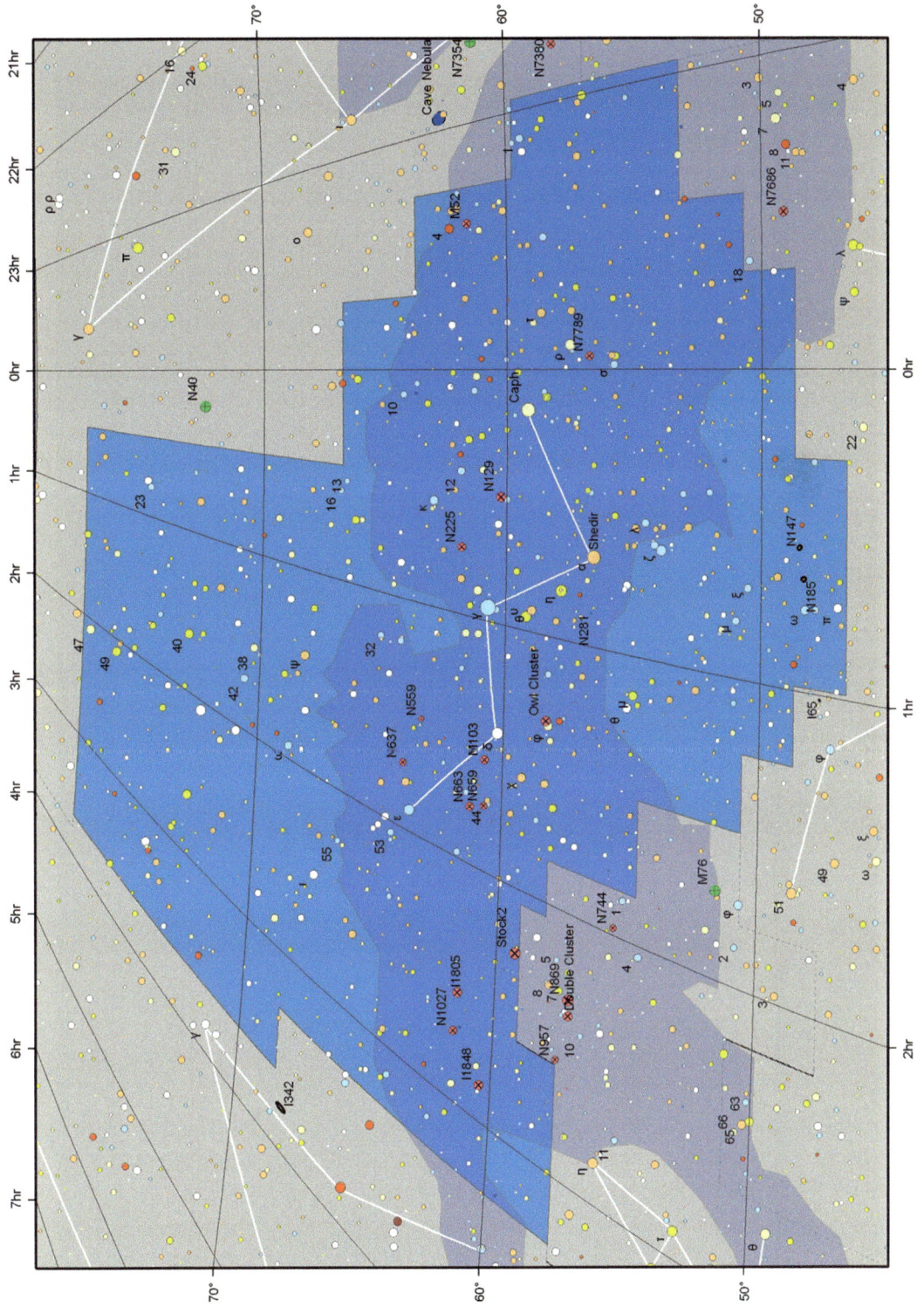

CENTAURUS

A very attractive southern hemisphere constellation located on the northern edge of the southern Milky Way. It's not the easiest constellation outline to memorize, but is very much worth the effort as there is a lot of good stuff to track down here.

The star alpha Centauri (also known as Rigel Kentaurus) is not only the brightest star in Centaurus and one of the brightest in the sky, but also a member of a triple star system containing the three stars closest to the Earth.

Historically

Centaurus is a large constellation, which, unsurprisingly, represents a centaur – a creature that is half man and half horse. Legend has it – no doubt suitably embellished by Ovid – that in this case the particular centaur represented is Chiron, but sources differ. Chiron, a child of Cronus, became a widely respected teacher who tutored heroes such as Achilles and Ajax, and also Asclepius – otherwise known as Ophiuchus.

Notable Stars

Nearby star: Proxima Centauri is the closest known star. It lies just 4.22 light years away from the Earth and is one of the most easily observed red dwarfs in the sky. Look for this variable 11th magnitude star just 2.2° away from Alpha Centauri.

Double: Alpha Centauri −0.0/1.3 and 1.7″. Two bright stars orbiting each other every 80 years. A third star, Proxima Centauri, orbits further out.

Double: Beta Centauri 0.6/4 and 1.4″. It is very difficult to separate these two blue stars.

Sunlike stars: Alpha and Beta Centauri are both sunlike stars. In color they are yellowish, as their composition and temperatures are similar to those of our Sun.

Deep Sky Objects

NGC5139: Omega Centauri is one of the skies must-see objects; far, far better than M13. A large, bright and easily resolved globular cluster, it can be spotted as a blurry 4th magnitude patch that is resolved into stars in a 100mm.

NGC3918: A bright and compact planetary nebula sometimes called the Blue Planetary (though color is rarely discernible visually in the deep sky). It is 8th magnitude and just 9 arc seconds across. The HST images of it are beautiful.

NGC5128A: Centaurus is a massive elliptical galaxy with a huge mass of dust in a ring round what appears to be its equator. An easy galaxy to spot. Its magnitude is 6.8 and it is 20 arc minutes wide: so it has a good surface brightness. The galaxy has a hugely active core, complete with supermassive black hole spewing jets out into space. Well worth seeking out.

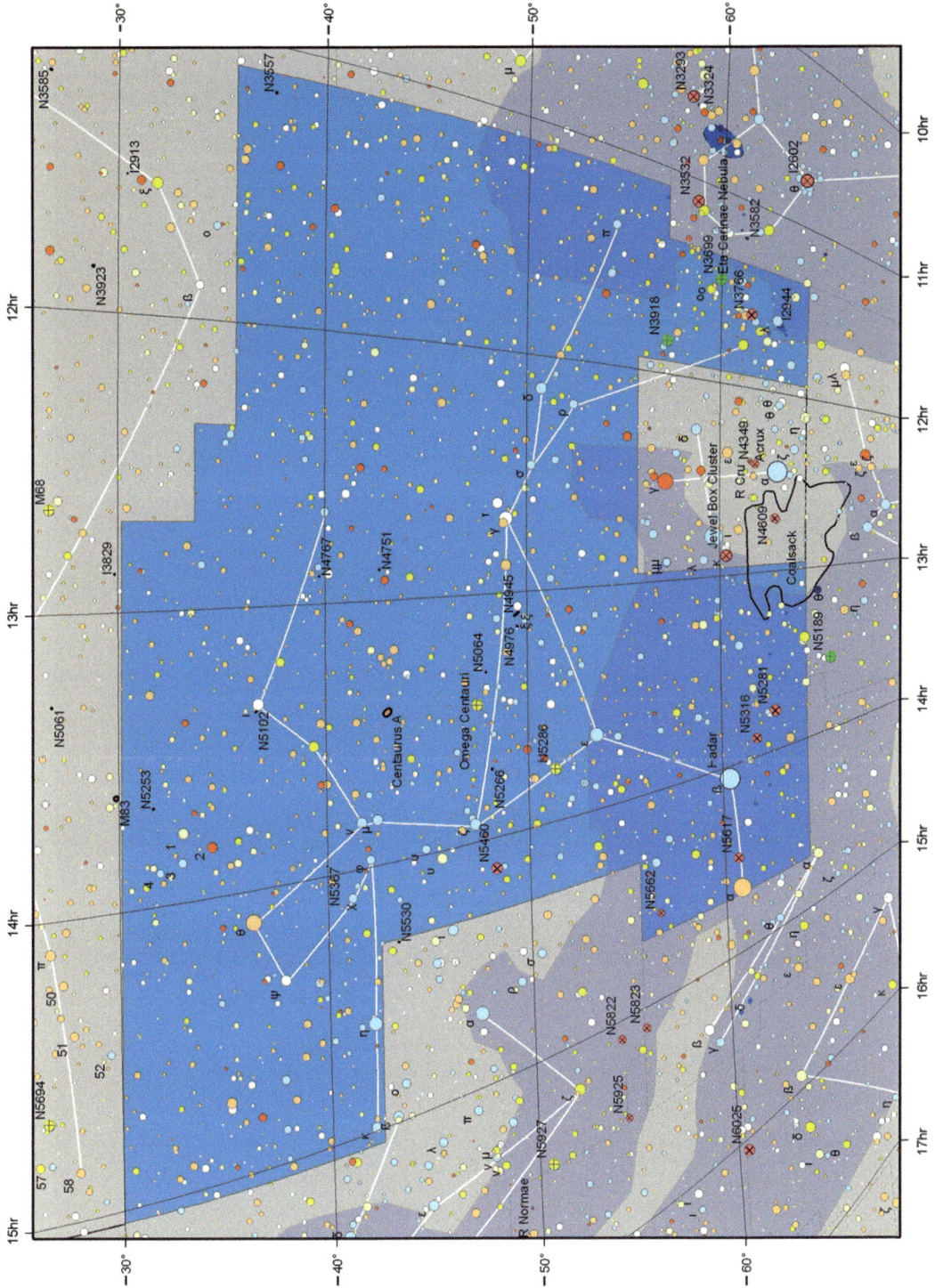

Fig. 9 Omega Centauri, the finest globular cluster in the sky: it resolves readily in small telescopes (Image by Bill Snyder and Chris Picking)

CEPHEUS

An attractive and distinctive northern hemisphere constellation stretching from Cygnus and Cassiopeia to near the celestial pole and Polaris. The Milky Way passes through the constellation, so several worthwhile nebulae and clusters lie within it and low power sweeping within its bounds is very worthwhile.

Historically

Greek mythology identifies Cepheus as the King of Ethiopia and husband of the vain, and not very bright, Queen Cassiopeia. Her daughter Andromeda is nearby in the sky, as is Perseus who is seen rescuing her from a sea monster – possibly represented by Cetus. Later Perseus and Andromeda were married and they all lived happily ever after apart from those slaughtered in the wanton carnage that broke out at the wedding – it's an entertaining story that Quentin Tarantino would be proud of.

Cepheus is one of the 48 constellations identified by Ptolemy.

Notable Stars

Double: Beta Cephei 3.5/7.9 and 13.3″. White and blue.

Double: Xi Cephei 4.5/6.5 and 7.8″. Blue and yellow.

Double: Delta Cephei 3.9/6.5 and 41″. Orangeish and blue.

Variable: The archetypal Cepheid variable star that has been used as a yard stick for measuring the distance of other galaxies because its luminosity and period are closely related. It varies between magnitude 3.5 and 4.4 in 5.36 days.

Deep Sky Objects

NGC7023: A lovely reflection nebula centered on a small cluster of stars, one of which is 7th magnitude. The nebula itself is 7th magnitude and is 8 arc minutes across. A 150mm aperture will be required to pick it out.

NGC6939: An open cluster of stars with a brightness of 8th magnitude. It contains about 100 stars in its 8 arc-minute diameter and its brightest member is about 12th magnitude. Stars form a straight line along the cluster edge.

NGC6946: A face-on spiral galaxy on the edge of Cepheus. It's dim at magnitude 11, but is in the same low power field as NGC6939. Challenging.

NGC188: A northerly open cluster. This 15 arc minutes wide, 9th magnitude cluster is very old and was shining billions of years before brighter newcomers like the Pleiades. It contains 50 stars of 12th magnitude or fainter.

NGC40: A nice planetary nebula, which is 10th magnitude and 5 arc seconds across. It's central star – a white dwarf – is 12th magnitude.

NGC7129: 4 or 5 faint stars (brightest magnitude 10) with nebulosity 5 arc minutes wide that will be visible in a 150mm aperture under dark skies.

Fig. 10 The fine spiral galaxy NGC6946 lies close to the Cepheus/Cygnus border and is seen in the same low power field as NGC6939 (Image by Grant Privett)

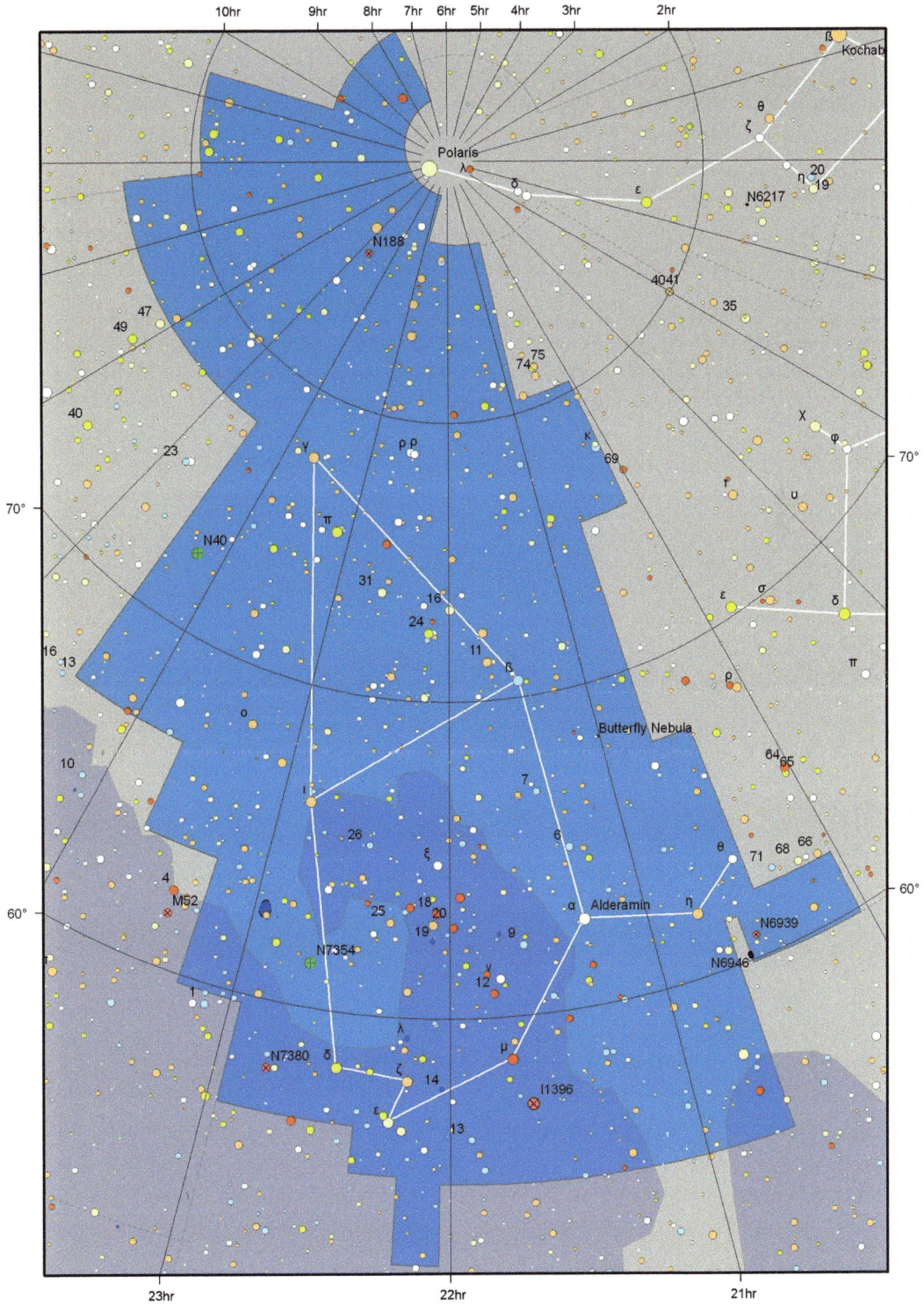

CETUS

Cetus is a constellation that encompasses part of the celestial equator. The majority of it lies to the south but the distinctive tail shape – including 2nd magnitude alpha Ceti – lies to the north. The constellation is easily spotted as it forms the outline of a cartoon whale shape. It is also host to one of the most famous variable stars.

At its best during November.

Historically

The mythology behind Cetus is quite varied. In some accounts it's a hideous ravening sea monster – perhaps a bit like a crocodile – while in others it's a normal whale. In the guise of the sea monster, it's involved in the Perseus and Andromeda story mentioned on the previously.

Notable Stars

Double: Gamma Ceti 3.5/6 and 2.6″. Yellow and blue. A 10th magnitude companion lies 14 arc minutes away.

Variable: Mira – "The wonderful" – a red giant star in the tail of Cetus, that varies between roughly magnitude 3.5 and 10. It was discovered in 1596 by the German astronomer/theologian David Fabricius who first believed it a nova. He was also one of the first Europeans to observe sunspots.

Sunlike star: Tau Ceti is one of the nearest sunlike stars. It lies 12 light years removed from us.

Deep Sky Objects

M77: A bright and easy to spot inclined barred spiral galaxy, with an active galactic core powered by a supermassive black hole. It's a Seyfert galaxy, bigger than the Milky Way, with a magnitude of 9. Amateur telescopes only show the central portion of the galaxy (2 arc minutes wide) while imagers see fainter spiral arms extending out to 8 arc minutes. M77 is visible in an 80mm refractor.

NGC245: A large and quite dim planetary nebula. It's about 8th magnitude and 4 arc minutes across – which spreads the light out. The central star is 12th magnitude.

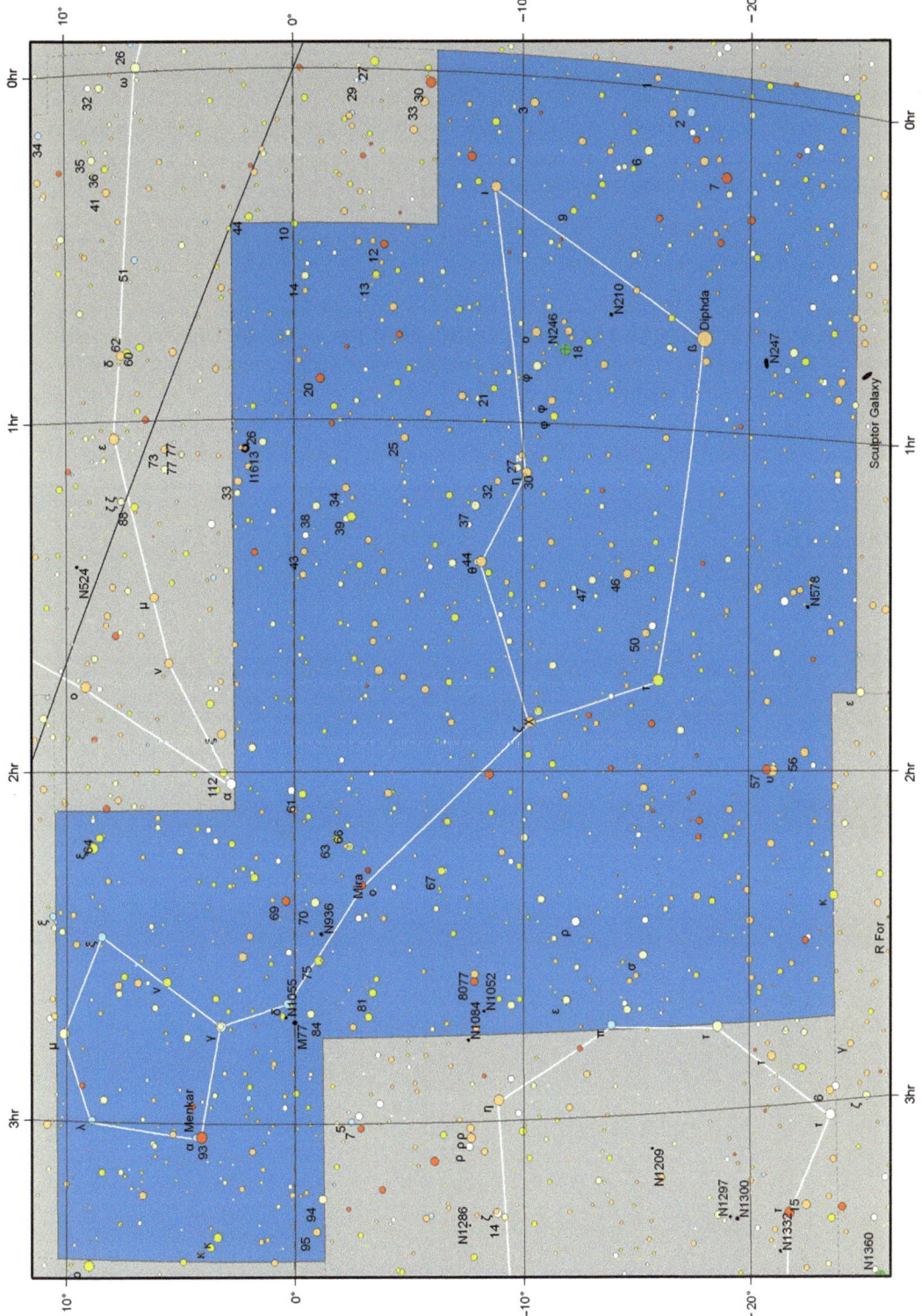

CHAMAELEON

A southern hemisphere constellation that hangs about 10° below the arc of the Milky Way, and above the South Celestial Pole as seen from the southern hemisphere, during the months of April and May.

This assemblage of rather dim stars – the brightest is 4ᵗʰ magnitude – apparently makes up the shape of a chameleon. I can't say I am convinced, but it is easy to spot. It lies on the edge of a fairly barren portion of sky and so has little competition.

Historically

The Chamaeleon is based upon the observations of the sailors Keyser and de Houtman who sailed on a ship to the East Indies in 1595, and so saw the southern sky. Unlike the mechanistic additions of de Lacaille, the constellations created by these two gentlemen are mainly animals – we saw the Bird of Paradise earlier. Unfortunately one of the sailors, Keyser, was killed on the voyage and his observations were eventually returned to Europe and to Petrus Plancius, who created star charts from them.

Notable Stars

No double stars.

Deep Sky Objects

NGC3195: The only deep sky object worthy of mention in Chamaeleon is an oval planetary nebula 35×40 arc seconds wide. It shines at 11ᵗʰ magnitude and so is visible in a 150mm telescope under good conditions. At a declination of −80, it's one of the most southerly planetaries.

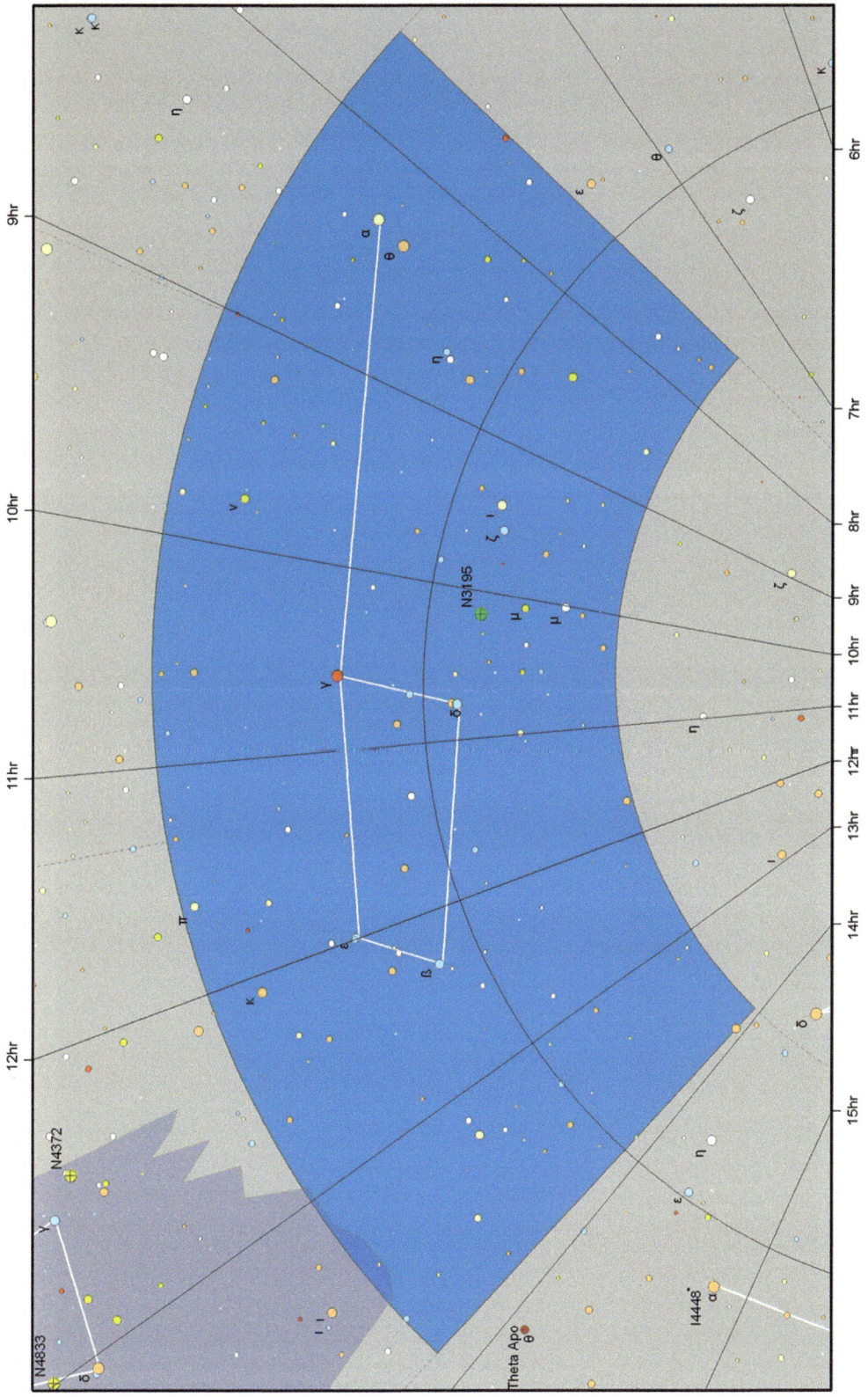

CIRCINUS

A southern hemisphere constellation created from a long and thin triangle of stars adjacent to, and east of, the brilliant alpha and beta Centauri. The stars are not dim, and Circinus lies within a fine portion of the Milky Way but, alas, has no notable deep sky objects to further recommend it.

Historically

Another constellation created by de Lacaille during his stint at the Cape of Good Hope cataloguing stars. Circinus represents a type of drawing compass used by draughtsmen, which is appropriate, given that at one time he was employed to survey the French coast between Nantes and Bayonne.

Notable Stars

Double: Alpha Circini 3.2/8.5 and 15.6″. Yellow and red stars.

Double: Gamma Circini 5.1/5.5 and 2″. Bluish and yellow stars.

Deep Sky Objects

There are no deep sky objects of note in Circinus, but it is in the Milky Way and so may be worth low-power sweeping.

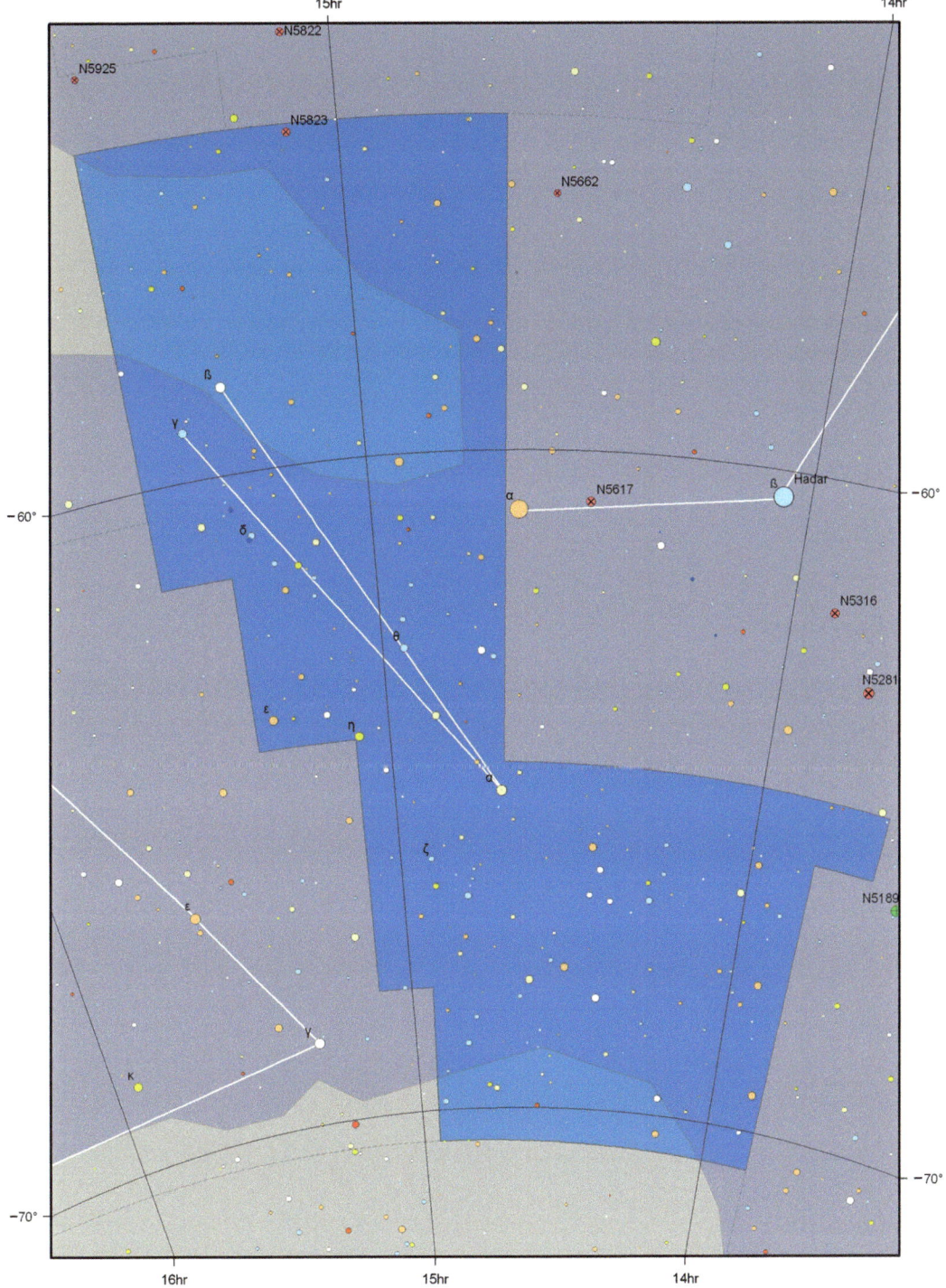

COLUMBA

A southern hemisphere constellation located immediately south of Orion and Lepus. Its stars are reasonably bright and the distinctive form easily spotted. Best viewed during February.

Historically

Another constellation first depicted by Petrus Plancius in the late sixteenth century. Originally he described it as being Noah's Dove – Petrus being a Minister of the Dutch church, who later became a cartographer – but it was eventually abbreviated to merely Dove. Not all the constellations invented by Plancius have survived to the modern day – included in the list of the abandoned are constellations depicting the River Jordan and the River Euphrates.

Notable Stars

No double stars

Deep Sky Objects

NGC1851: This globular cluster in Columba is it's only deep sky object. A nicely condensed 7th magnitude some 8 arc minutes across. Its brightest stars are 13th magnitude. The center of the cluster is sharply brighter.

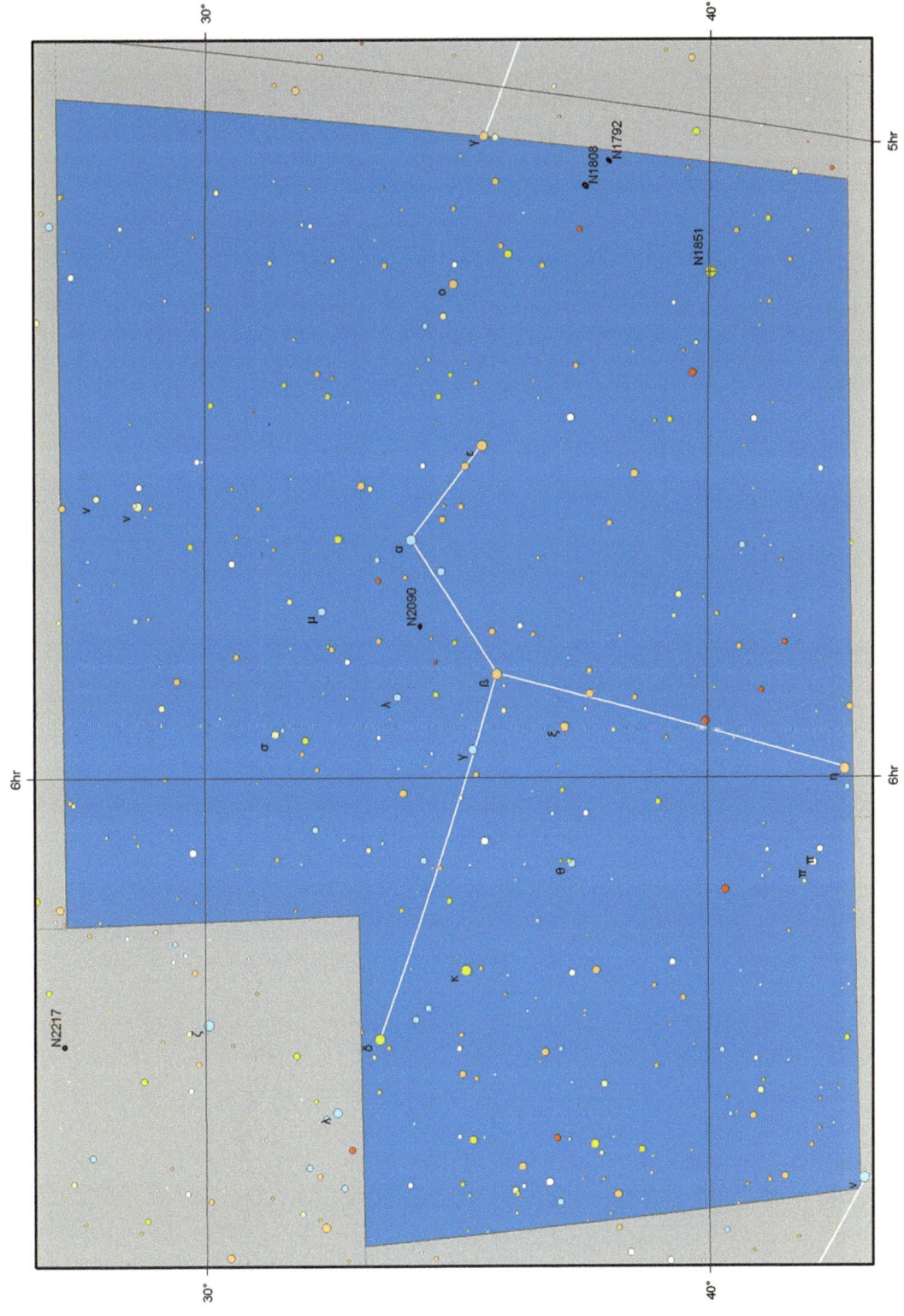

COMA BERENICES

A loose speckling of dim stars south of Canes Venatici. The brightest star is 4th magnitude, with many of the others being members of the large Melotte 111 open cluster – it actually has 50 or so members brighter than 10th magnitude.

Coma is home to a substantial galaxy cluster and is quite rich in deep sky objects, especially galaxies.

Historically

An old constellation formed from a grouping of dim stars. It doesn't appear in Ptolemy's list, though he noted its existence. A Greek poem refers to the hair of Berenice being cut off as an offering to the gods, so that her husband should return safe from war – he did, and hence the hair is now seen in the sky. Some Greek writers also referred to it as Ariadne's Hair.

Notable Stars

No double stars

Deep Sky Objects

Melotte111: The symbol shown indicates the centre of this large sprawling cluster. It makes up much of the western part of the constellation and is spread over 5–8°, forming a vaguely triangular shape. An impressive sight in binoculars.

M64: The Black Eye Galaxy. An inclined barred spiral galaxy with tightly wound arms that has, superimposed upon it, a distinct dark patch caused by a dust cloud. The galaxy itself is 8th magnitude and it is discernible in binoculars. A 150mm aperture will be needed to detect the black patch near the galaxys' core. It will appear broadly 9 arc minutes by 4 arc minutes wide; in a 200mm the feature is obvious.

M53: A loosely packed, but worthwhile, 7th magnitude globular cluster not far from alpha Comae Berenices. It can be seen in binoculars as a dim hazy star or – technical term – a blob. In a 100mm aperture the cluster will start to be resolved – first becoming clumpy or granular, and then more obvious as the aperture increases in size. The cluster is 5 arc minutes across.

M85: A 9th magnitude lenticular galaxy that will appear broadly 3 arc minutes across. Next to it is a faint barred spiral galaxy.

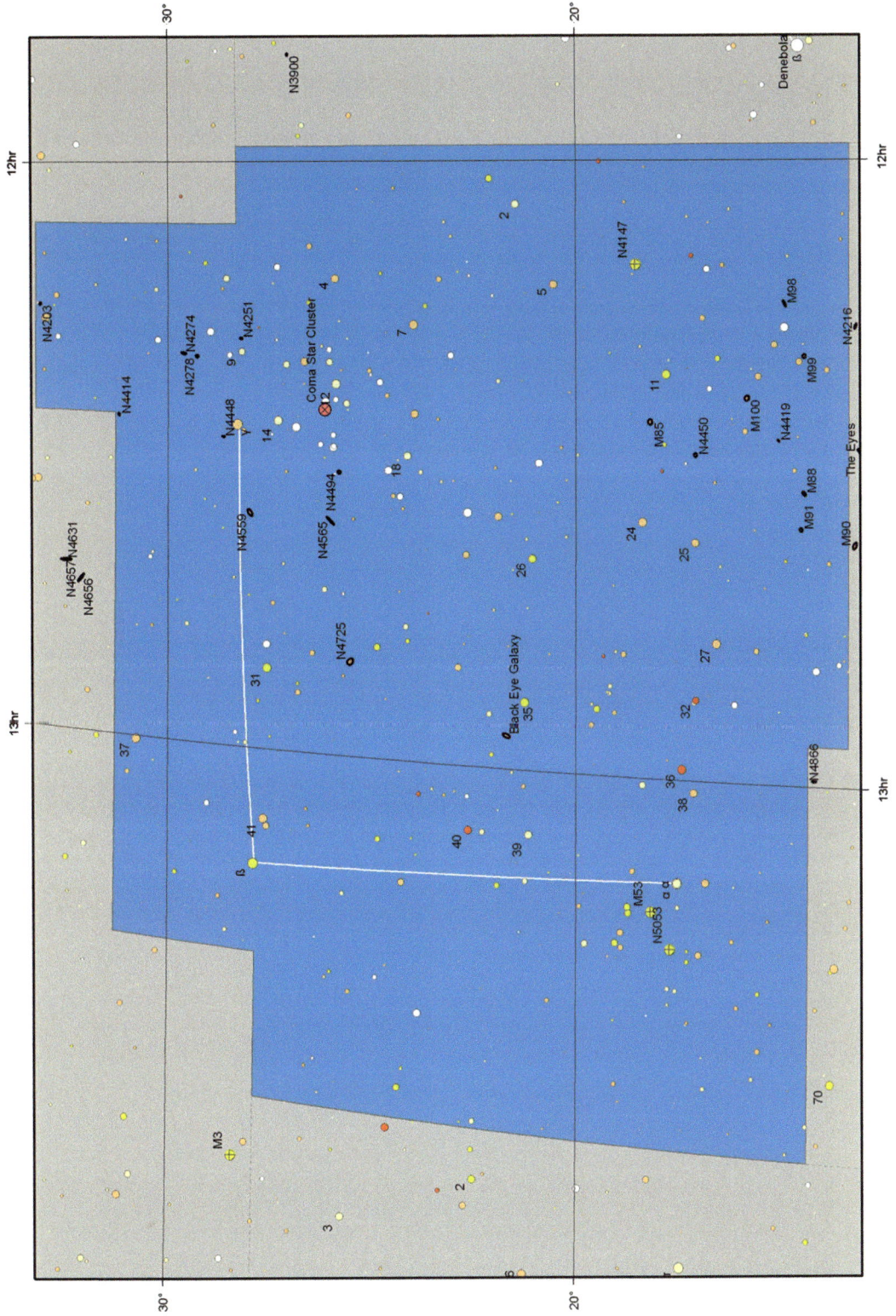

CORONA AUSTRALIS

A southern hemisphere constellation just south of Sagittarius and east of Scorpius. The stars are not especially bright – at best, 4th magnitude – but their closeness and the curve they make does render it quite easy to spot Corona Australis for the first time. Nestling on the edge of the Milky Way, the low power sweeping is worthwhile.

Historically

Corona Australis, the southern crown, appears in the second century book the *Almagest*, produced by the astronomer Ptolemy. The *Almagest* was a fairly comprehensive summary of what was known or surmised (the Greeks were rather prone to guessing without evidence) about the stars, planets, earth and universe at the time.

On charts Corona Australis often appears as a wreath rather than a crown. None of the stars are very bright, but it is still quite obvious.

Notable Stars

Double: Gamma Corona Australis 4.9/5.0 and 1.5″. White and white. Needs a high magnification.

Double: Kappa Corona Australis 5.7/6.3 and 21″. A line of sight double. Blue and white.

Deep Sky Objects

NGC6541: The only deep sky object in Corona Australis is this 6th magnitude globular cluster. It should be visible in large binoculars and is certainly visible in an 80mm refractor. It will appear between 3 and 10 arc minutes across depending upon your telescope, but it will take a 250mm aperture to resolve.

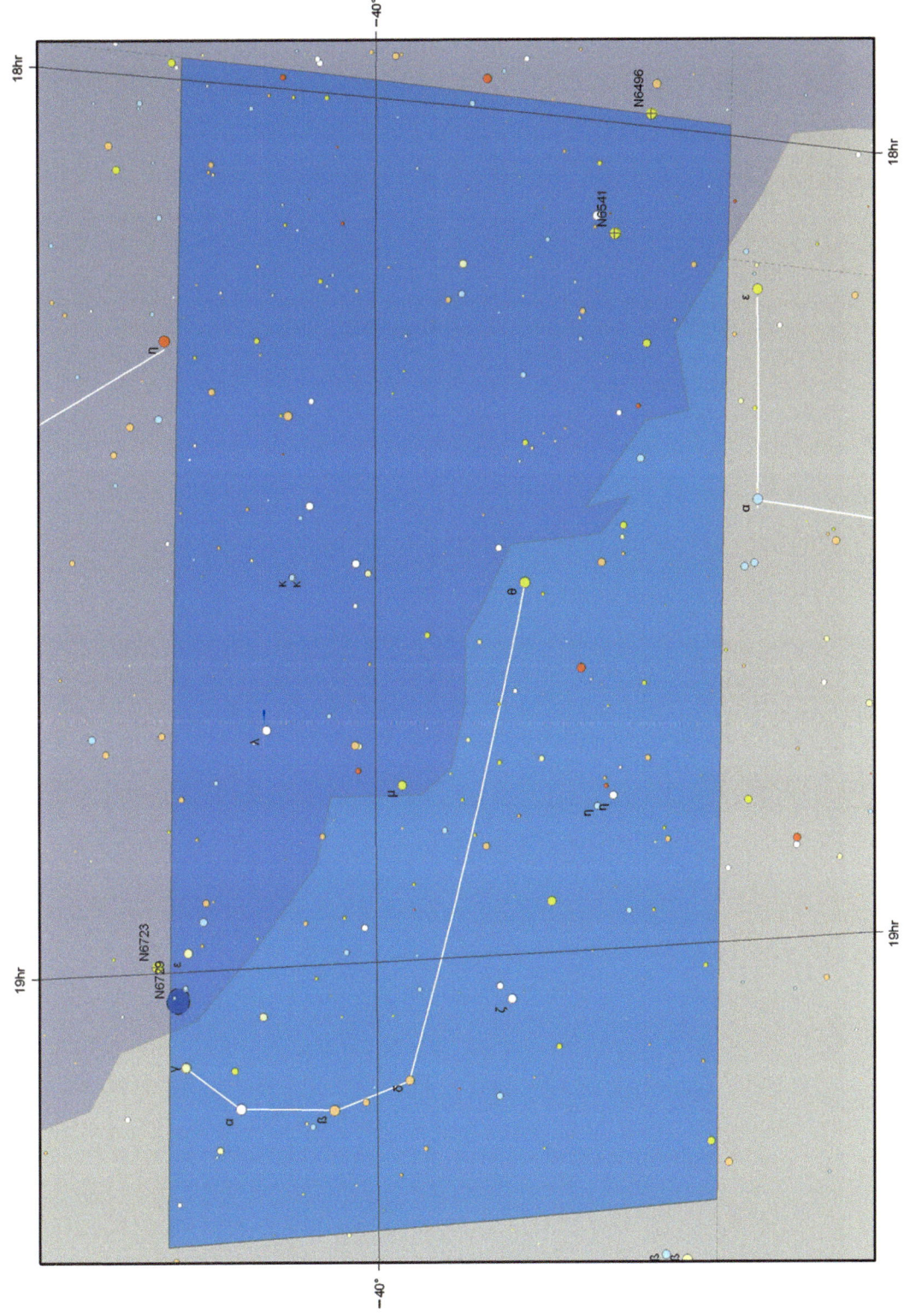

CORONA BOREALIS

A northern hemisphere constellation and one of the few that look something like the item it is alleged to represent. Its location next to Bootes makes locating it trivial. At its best during July.

Historically

The shape of the constellation, a curved arc of stars, inevitably led to the identification with a crown in several cultures. The particular crown it refers to depends upon the culture being discussed. However, the indigenous Australian people, see the curved shape as representing a "Woomera", a curved throwing weapon – a name employed for the Australian government research site in Australia that was used to launch rockets.

Its location, in the northern hemisphere of the sky, explains the use of the term borealis. It is one of the constellations recorded by Ptolemy.

Notable Stars

Double: Zeta Coronae Borealis 5.1/6.0 and 6.3″. Bluish stars.

Double: Sigma Coronae Borealis 5.0/6.0 and 6.2″. Yellow stars.

Variable: R Coronae Borealis fluctuates between magnitude 5.7 and 14 irregularly, probably due to obscuring clouds of carbon dust forming in its atmosphere. It was discovered by Edward Piggot, who also discovered Eta Aquilae.

Deep Sky Objects

There are no notable deep sky objects in Corona Borealis.

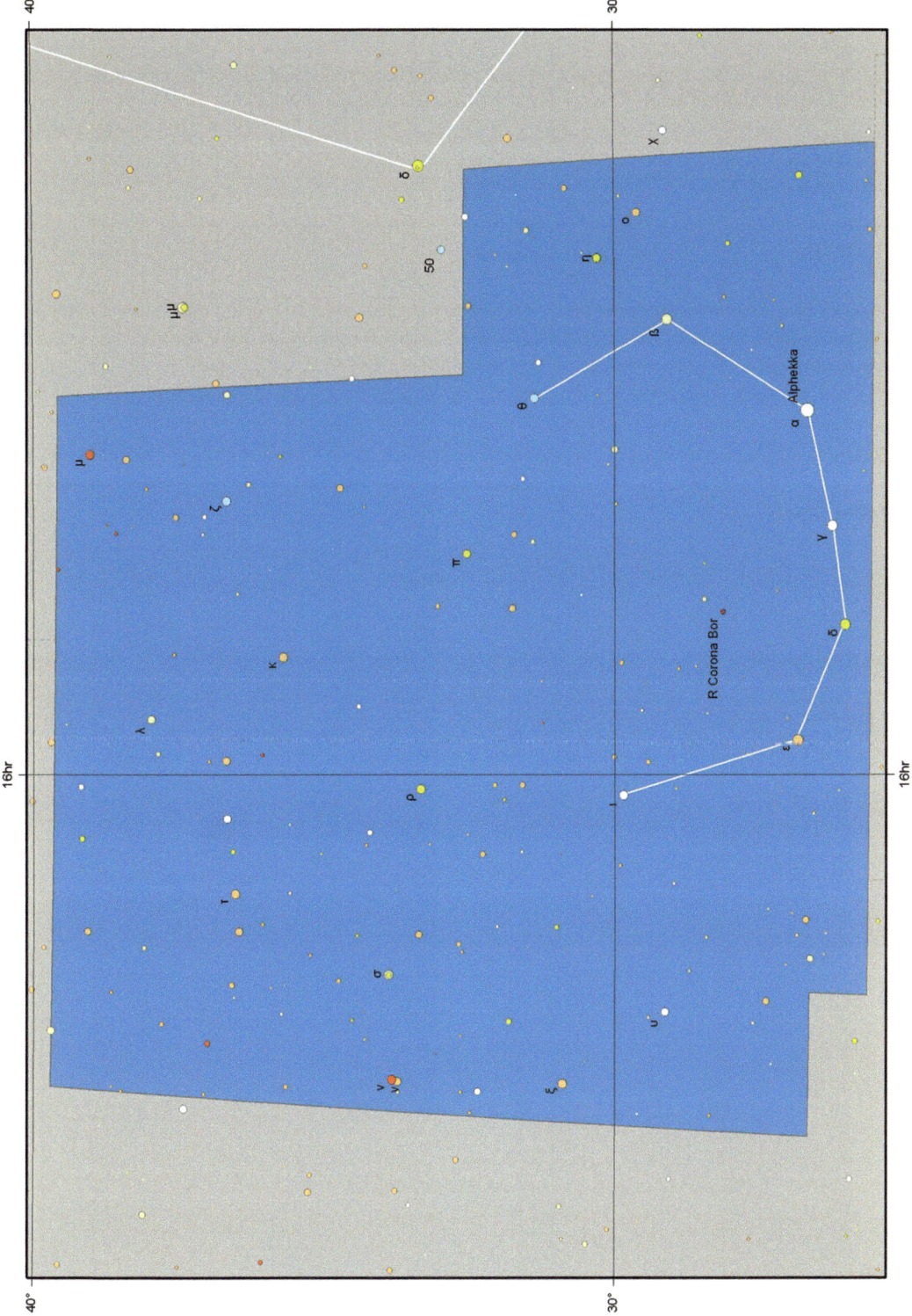

CORVUS

Corvus is an easily spotted, and surprisingly memorable, rhomboid shape of 2nd and 3rd magnitude stars located just above Hydra and to the west of the brilliant star Spica – alpha Virginis. It lies in the southern hemisphere of the sky, but not so far south as to be unseen in Europe and north America. Best located during May.

Historically

An ancient constellation representing a crow, which was asked by Apollo to collect some water. The nearby constellation Crater, is the cup the crow eventually brought some water back in. However, the crow got seriously side-tracked by some tasty looking fruit and so Apollo, pretty fed up with waiting, eventually cast the crow and the cup into the sky (as you do). They are separated slightly by a water snake – Hydra.

Notable Stars

Double: Delta Corvi 2.9/8.5 and 24″. Blue-white and reddish.

Deep Sky Objects

NGC4361: A dim planetary nebula is the only worthwhile deep sky object in Corvus. Will be seen in a 150mm, but is 10th magnitude and 1.2 arc minute across.

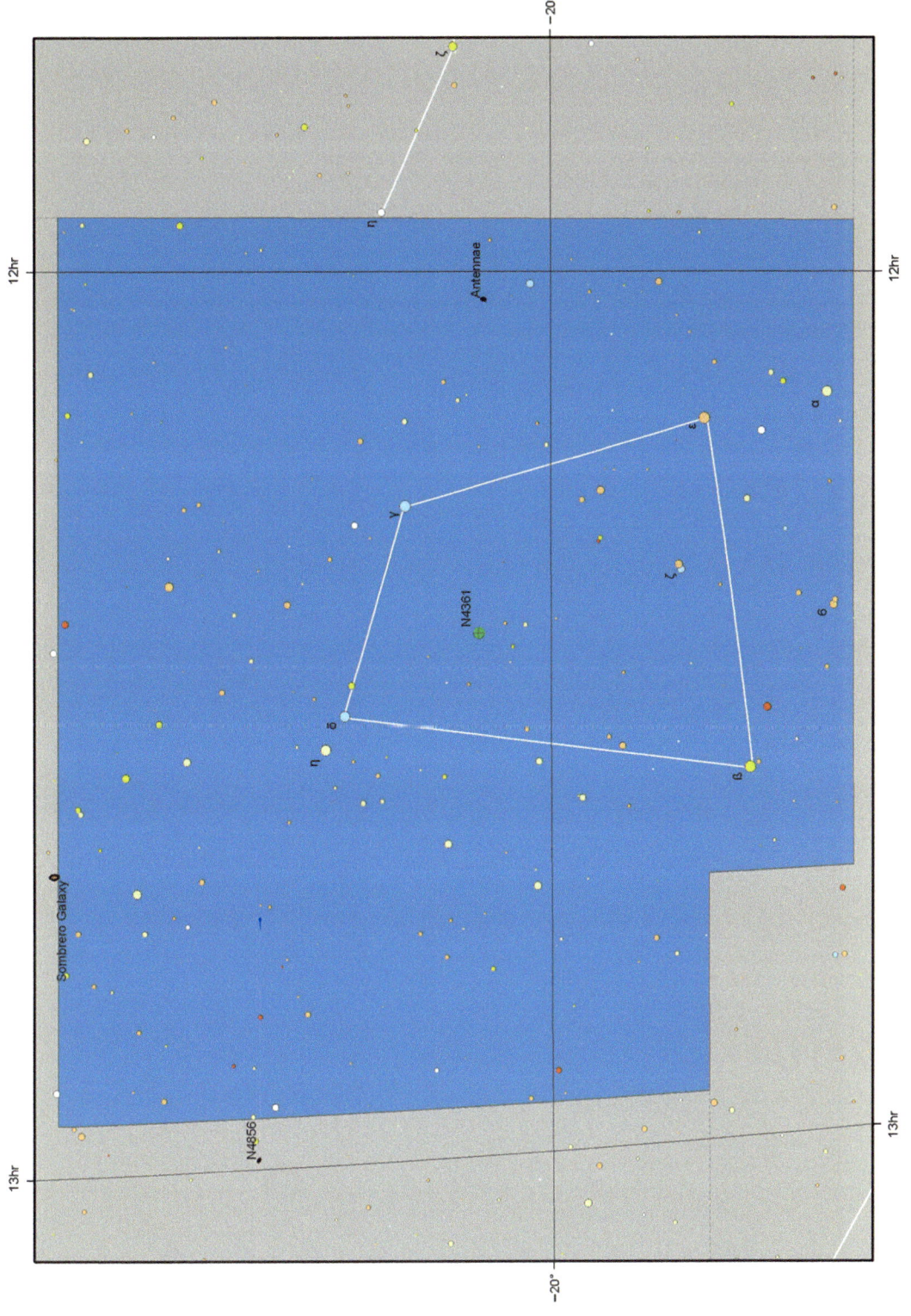

CRATER

A curve of stars lying on the back of the sprawling shape of the giant Hydra. It does look a bit like a cup but two of its stars, alpha and beta, would equally well have fitted into the shape of Hydra. Its brightest star is the 3rd magnitude star delta Crateris.

A southern hemisphere constellation well viewed during April.

Historically

Crater is an ancient constellation and in mythology associated with Corvus, the crow, described on the previous page under Corvus.

Notable Stars

Double: Gamma Crateris 5.0/5.0 and 2.7″. Very even white stars.

Deep Sky Objects

There are no notable deep sky objects in Crater.

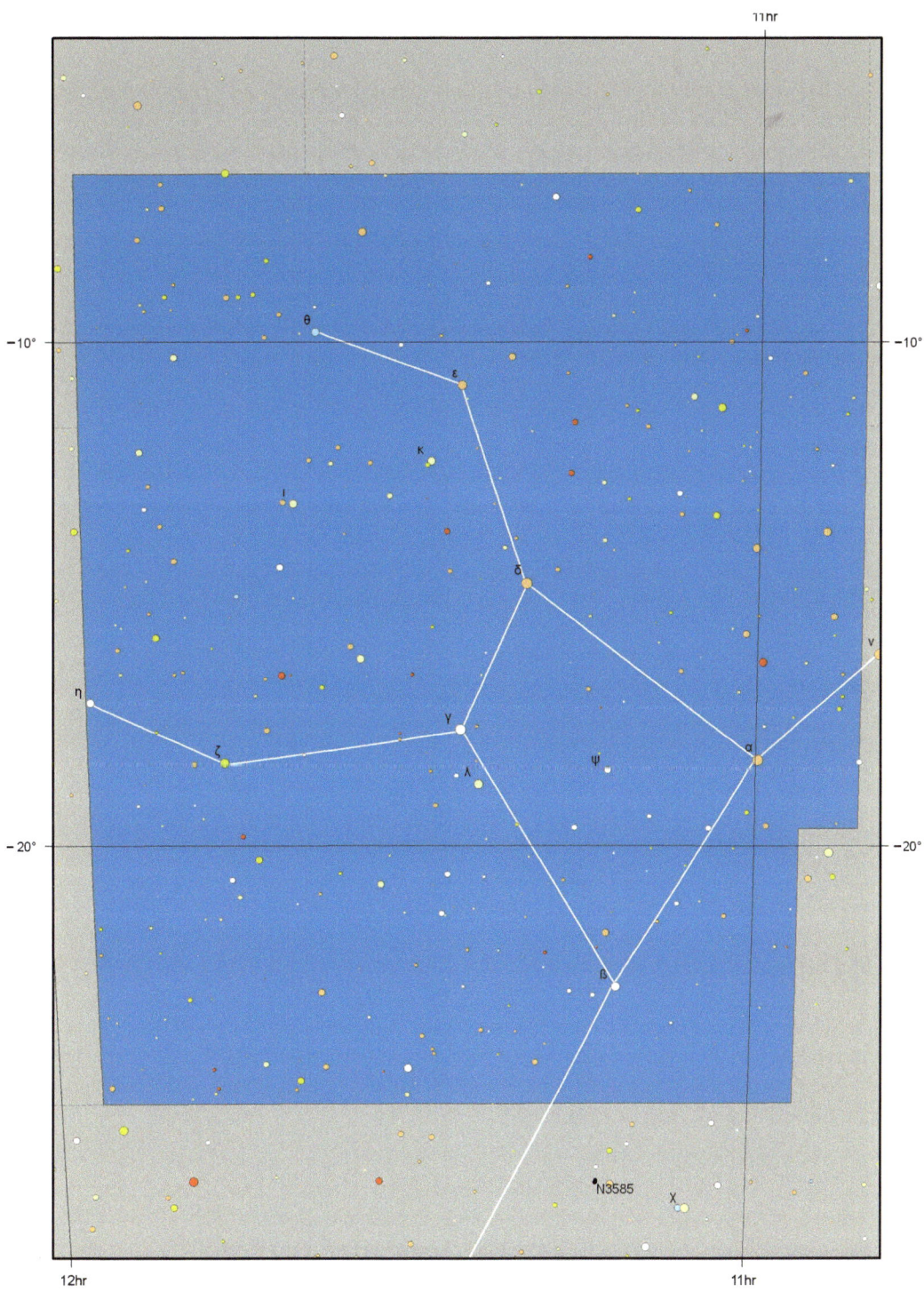

CRUX

An iconic southern hemisphere constellation best seen during May and June. It lies in a particularly brilliant region of the Milky Way which is strewn with bright stars. Because of its distinct cross-like shape it is very easily spotted even by someone visiting the southern hemisphere for the first time. It provides a great starting point for lower power sweeping with a telescope, or browsing with binoculars.

It's the smallest of the constellations, covering a mere 68 square degrees from the skies complete area of 41,200 square degrees. Despite that, its four main stars are quite bright. Crux is found on the flags of New Zealand and Australia and is generally known as the southern cross – although precession means that a mere 4,000 years ago its shape was visible from the 50N latitude, and it will be again, in the future.

Historically

The southern cross was once visible from Europe and Egypt but has gradually drifted south and so was unknown to the Romans and Greeks. For a while it was considered part of Centaurus – as Ptolemy had recorded – but is now viewed as a separate, if small, constellation in its own right. As you might expect, cultures differ in what they see in its form, with some perceiving it as a small bird and others as a fish.

Notable Stars

Double: Alpha Crucis 1.3/1.6/4.8 and 4.0/90″. A triple star. All stars blue.

Double: Gamma Crucis 1.8/6.5 and 125″. Red and white. Line of sight double.

Double: Mu Crucis 4.0/5.1 and 35″. Blue stars.

Variable: R Crucis a cepheid variable oscillating between magnitude 6.4 and 7.2 in 5.8 days.

Deep Sky Objects

Coalsack: A large and obvious dark nebula that can be seen silhouetted against the Milky Way beyond. It can be detected from semi-urban locations under good conditions, but is best viewed from a dark site where it appears to consist of two parts, covering more than 6–7° of sky on the southwestern edge of Crux and spilling into Centaurus and Musca.

NGC4755: The beautiful Jewel Box cluster. It is an attractive sight in small telescopes and dominated by the blue super-giant, kappa. A brightish red star lies a little off from the cluster heart, making a strong and attractive contrast with the other stars. Many of the cluster stars appear to form short straight runs of 3–4 stars. The cluster is a compact 10 arc minutes across and contains 50 stars.

NGC4609: A small cluster 5 arc minutes wide containing 50 stars. Together they shine at magnitude 7. It probably lies beyond the Coalsack.

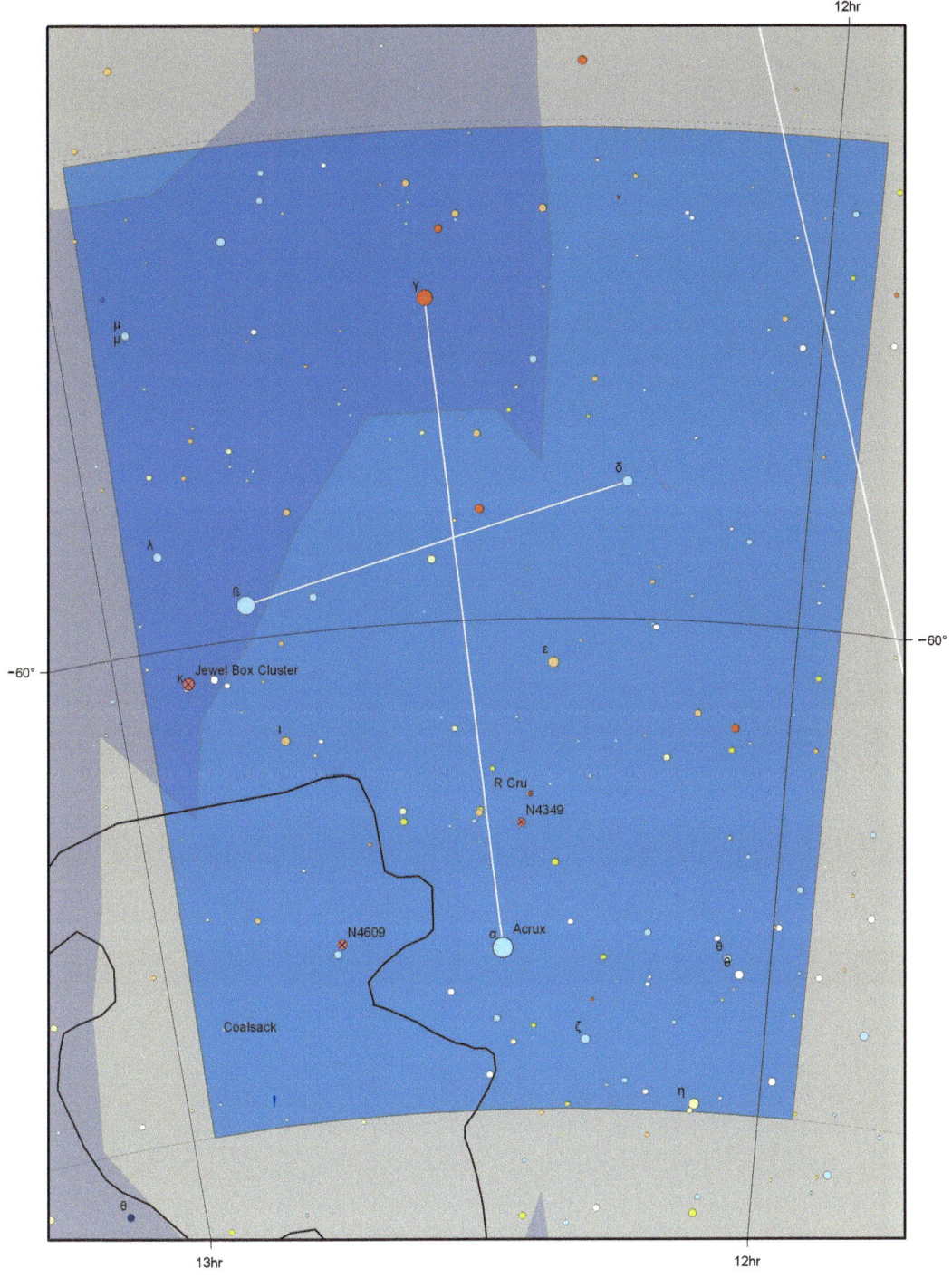

Fig. 11 The Coalsack is a dark nebula on the edge of Crux. It stands out against the bright Milky Way beyond (Image by Grant Privett)

CYGNUS

A famous and familiar form of the northern hemisphere sky. This large cruciform really does bring to mind a swan, its neck outstretched, flying overhead. It's quite a large constellation and lies right on top of some of the best parts of the northern hemisphere Milky Way.

On a dark night the Great Rift – a region of the Milky Way attenuated by dust and gas – runs north–south, from the brilliant Deneb at the swans tail down to the beautiful double star Alberio at its head. A must see constellation.

Historically

The distinct cross shape deep in the Milky Way represents the swan that Zeus turned himself into when chasing down and assaulting a nymph, Leda. Zeus must rank high on the list of gods you would not want to meet in a dark alley. Other mythological figures were also turned, or turned themselves into swans: not least Orpheus. They made their own entertainment in those days.

Notable Stars

Double: Beta Cygni 3.1/5.1 and 34″. Yellow and blue stars. A truly lovely sight.

Double: Omicron Cygni 4/7 and 107″. Yellow and blue stars.

Double: Delta Cygni 3.0/6.5 and 2.2″. Blue and yellow stars.

Nearby star: The double star 61 Cygni is the 14th nearest star. The light we see leaving it has been on it's way for just 11.4 years. The stars are separated by 31″

Deep Sky Objects

M29: One of the less impressive Messier objects. A handful of 8th–10th magnitude stars enclosed within 10 arc minutes, lying on the edge of the Great Rift. There are seven brightish stars forming a truncated Pleiades-like shape. Images show the number of background stars dropping off as you cross the cluster and the obscuring gas and dust hides them.

NGC6826: The Blinking nebula. A planetary nebula of 8th magnitude covering an oval 24×27 arc seconds. It doesn't actually vary in brightness, but at certain power/aperture combinations the central star (magnitude 11) appears to brighten and fade as your eyes switch from peripheral to direct vision perception. Well worth a visit.

M39: A bright and loose cluster, easily visible in binoculars, covering half a degree. From truly dark sites it is visible to the naked eye. Its 30 or so brightest members will be well seen in a 114mm aperture Newtonian.

NGC7000: A cloud of nebulosity a degree across not far from Deneb. It's not particularly bright but is visible in 10×50 binoculars from dark sites. You will need to use binoculars or a wide field of view telescope to catch this against the surrounding darker sky. The name derives from its apparent shape.

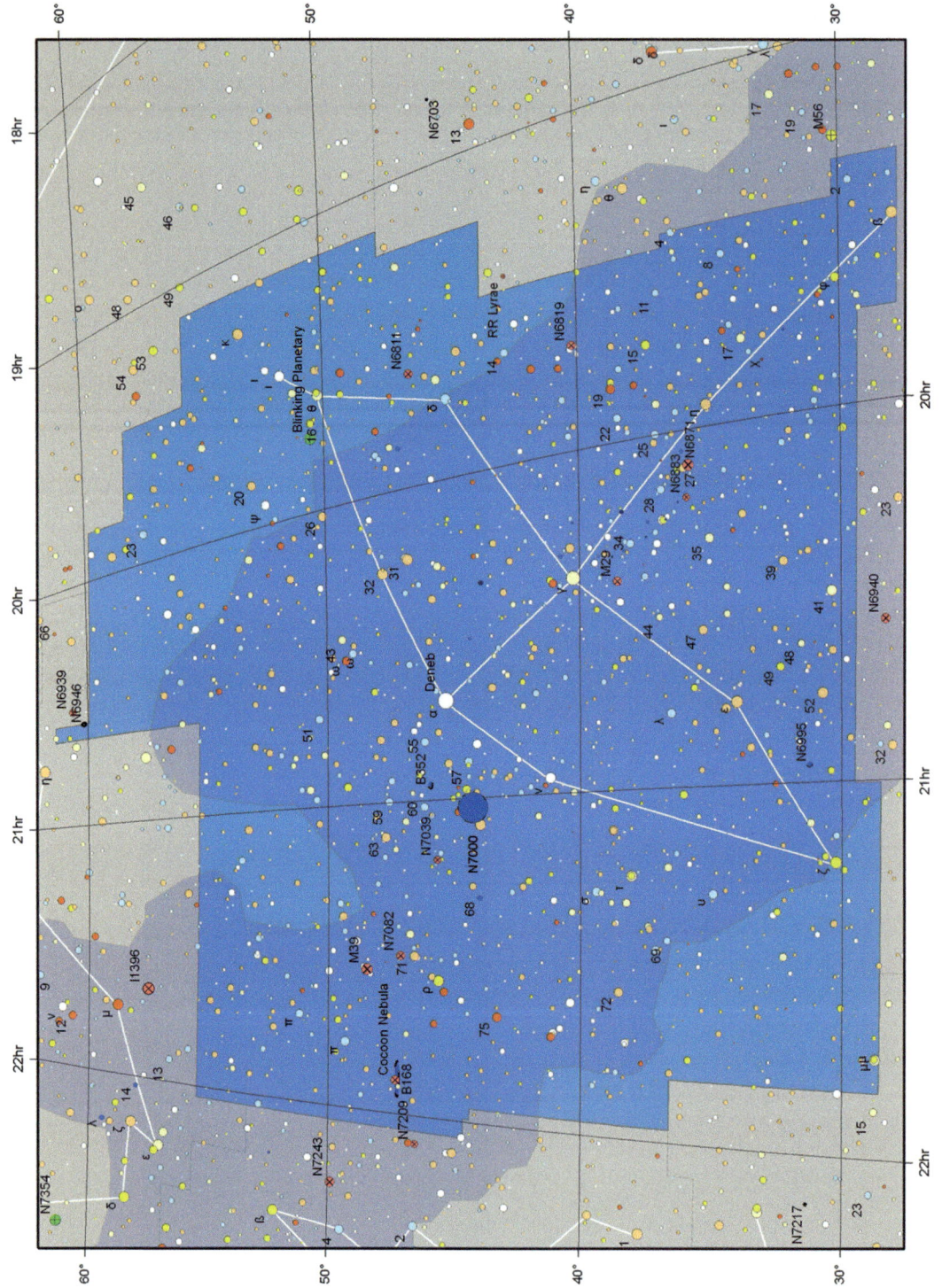

Fig. 12 The Cocoon nebula in Cygnus contains much detail and is detectable in a 114mm aperture (Image by Grant Privett)

DELPHINUS

A northern hemisphere constellation on the edge of the Milky Way. It's easily spotted under suburban skies and, with the equally striking Sagitta, can be used as a guide to locating the fainter neighboring constellations of Vulpecula and Equuleus. Its stars are of 3rd and 4th magnitude. Most easily located during October.

Historically

This small and easily memorized grouping of stars depicts a dolphin. In one story it commemorates the rescue of Arion – a great player of the Lyre (a stringed musical instrument) – who was carried across the sea on the back of a dolphin, away from some thieves. Alternatively, it was an ambassador of Poseidon who, via clumsy wooing, had frightened someone he wanted to marry.

Notable Stars

Double: Gamma Delphini 4.5/5.0 and 10″. Yellowish stars.

Deep Sky Objects

NGC6934: A 9th magnitude globular cluster that can be readily seen in an 80mm aperture. It appears roughly 2 arc minutes across and will probably not be resolved in an aperture less than 300mm.

NGC7006: Another globular cluster and one of the dimmest clusters presented in this book. It is notable for also being one of the most remote examples – it lies 130,000 light years away. It will be challenging for smaller instruments, as it shines at 10th magnitude, appears 2 arc minutes across and will require care and attention to discern. Unless you know someone with a 500mm Dobsonian, don't expect to see it resolved it into stars.

NGC6905: Magnitude listings often show this nebula as 11th or 12th magnitude, but observers often report it as being easier to spot than expected. See what you think. It is an oval, roughly 30×45 arc seconds across.

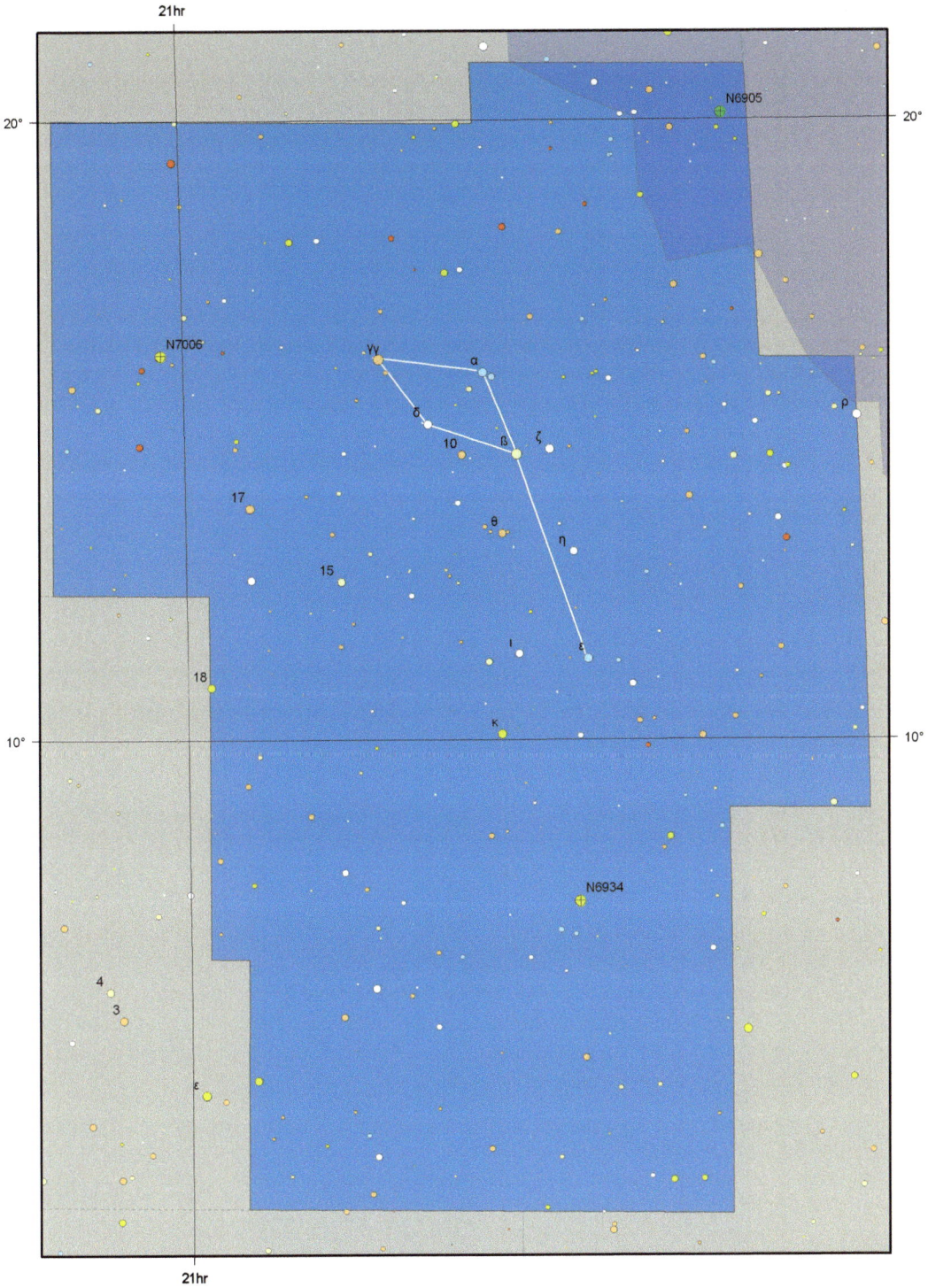

DORADO

A southern hemisphere constellation formed from a line of stars separating Reticulum from Pictor. Its brightest star, alpha, is 3rd magnitude and so Dorado might be considered unexciting were it not for the presence of the Large Magellanic Cloud – a satellite galaxy of our own Milky Way within its bounds. The deep sky objects associated with the LMC make the constellation well worth visiting. Look out for it at the start of the year.

Historically

Yet another creation of Petrus Plancius, whose name in his native tongue was Pieter Platevoet: Peter Flatfoot, in English. In those days, scholars communicated mainly in Greek or Latin, so his name was translated. It's worth mentioning that Plancius asked Keyser to collect the data needed for southern star maps – none were available at the time.

Dorado is unusual in being derived neither from Greek or Latin but is the Spanish for a type of fish. Some charts, including those belonging to Kepler, also refer to it as a swordfish, although this never caught on.

Notable Stars

No double stars.

Variable: Beta Doradus is a delta Cepheid variable star which oscillates between magnitude 3.8 and 4.7 in a bit under 10 days as its spectral type changes.

Variable: S Doradus lies in the Large Magellanic Cloud and is a brilliant blue supergiant. It is prone to outbursts. With a mass of more than 50 Suns it's a million times more luminous than our Sun.

Deep Sky Objects

LMC: The Large Magellanic Cloud. A small galaxy orbiting our own. It covers a chunk of sky 11×9° and is what remains of a barred spiral structure after several distorting interactions with our Milky Way galaxy. The LMC is 150,000 light years away and obvious to the naked eye – even moonlight doesn't always hide it. In 1987 it was host to a supernova explosion and contains several hundred clusters of its own. An unforgettable sight in binoculars and telescopes.

NGC2070: The Tarantula nebula. A massive star forming region within the LMC, this is a vast cloud of gas and dust full of intricate detail. The apparent size will vary depending upon the instrument used. Look for a glow 10 arc minutes or more across. Formerly known as 30 Doradus, a small cluster is associated with it. A fascinating object that dwarfs the nearer M42 in Orion.

NGC1755, NGC1866, NGC2004: The LMC contains a number of clusters – which are visible in telescopes of 150mm aperture. They are 9th magnitude or fainter and appear as nebulous patches just a few arc seconds across.

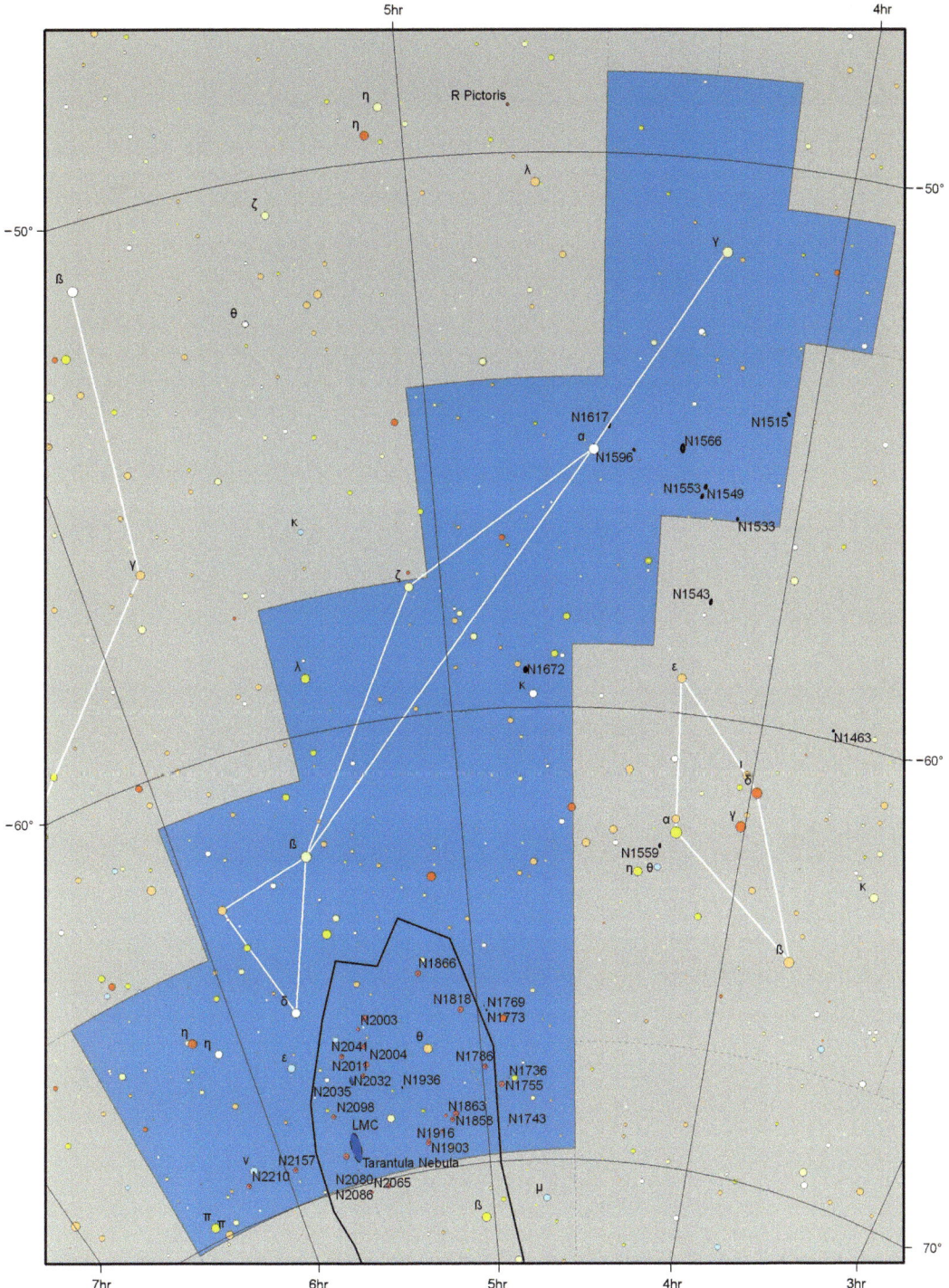

Fig. 13 The Large Magellanic Cloud in Dorado is a satellite galaxy of the Milky Way, which is larger and also brighter than the Milky Way. An easy sight in binoculars (Image by Chris Picking)

DRACO

A northern hemisphere constellation sprawling across 9 hours of right ascension and more than 40° of declination, not far from the North Celestial Pole. The sweeping tail nearly encompasses the constellation of Ursa Minor.

Draco sometimes produces an intense meteor shower centered on October 9th. Unfortunately, a bit like the Leonids, this is only happens a few times a century. Most years you should expect 10–20 meteors an hour under ideal conditions – with the radiant overhead and deeply dark skies – which will appear to radiate from near the dragons' head.

A constellation best placed during July.

Historically

It's no surprise that Draco is, apparently, a dragon or serpent strewn across the sky and around the celestial pole. As Draco lies adjacent to Hercules, the association with his 12 Labors is strongly supported. Apparently, the dragon was set to protect a tree that grew golden apples. Unfortunately, Hercules was required to pick apples from the tree. In the conflict that followed, the dragon was slain by Hercules and subsequently lofted into the sky by its owner, Hera, the wife of Zeus.

Notable Stars

Double: Psi Draconis 4.9/6.1 and 30.3″. Yellow stars.

Double: Nu Draconis 4.9/4.9 and 62″. Evenly matched white stars.

Double: Omicron Draconis 4.8/7.8 and 34.2″. Yellow and blues stars.

Double: Epsilon Draconis 4.0/6.9 and 3.2″. Challenging, but nice color contrast.

Sunlike star: Sigma Draconis is a sunlike star. It is 4.7 magnitude and fairly unimpressive – just as our Sun would look as seen from Sigma.

Deep Sky Objects

NGC6543: The Cat's Eye nebula. A fine, bright (8th magnitude) planetary made famous by the striking and colorful HST images of it. Some people do see hints of color (typically green blue rather than red) as the surface brightness is quite high. It's worth tracking down. Look for an oval 15×25 arc seconds across. The central star is 9th magnitude.

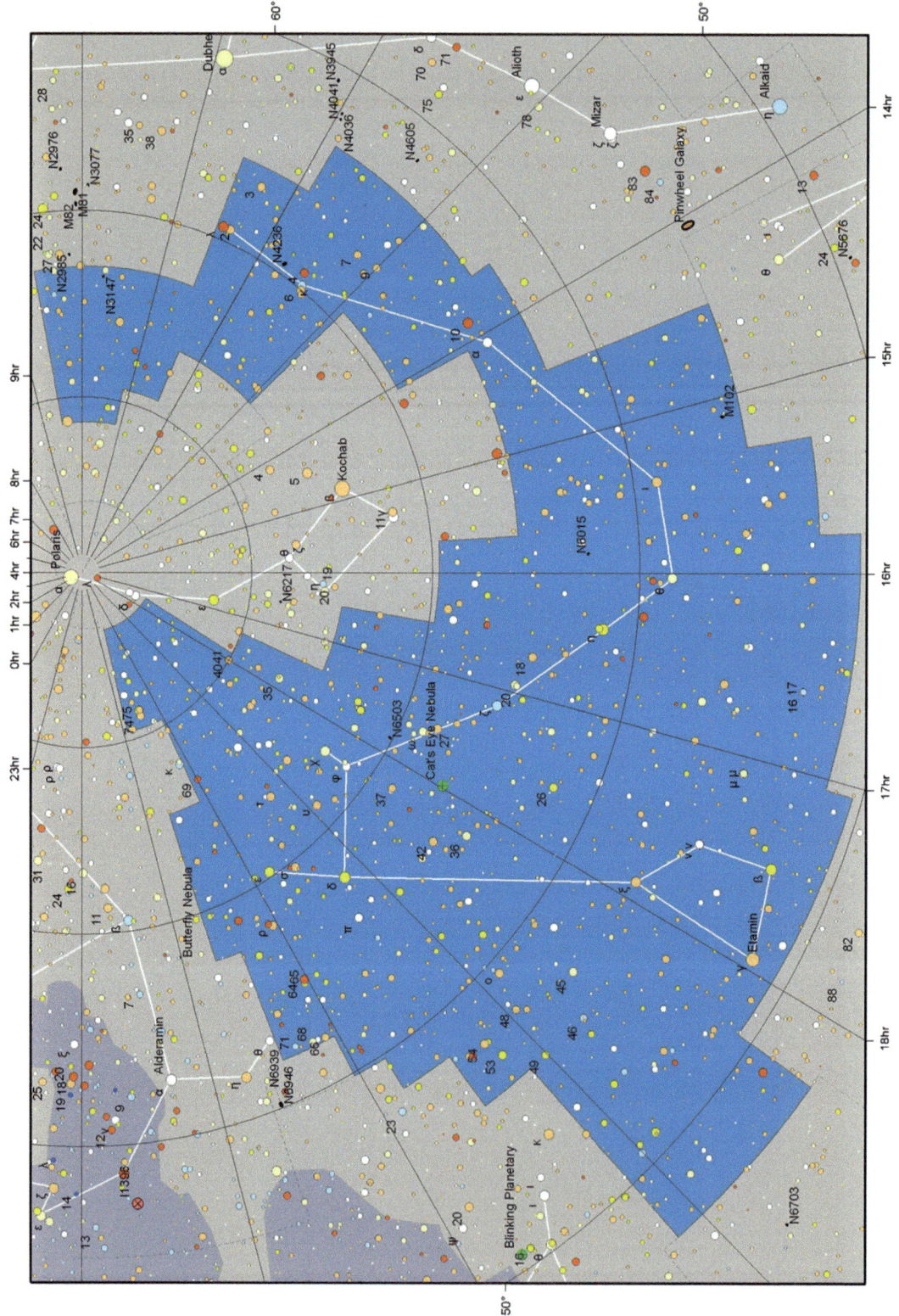

EQUULEUS

Equuleus is a singularly unimpressive and easily overlooked constellation that is the second smallest, covering 72 square degrees – beaten only by Crux which covers a paltry 68. The slightly larger Delphinus is far more noteworthy.

This northern hemisphere constellation is bounded by Delphinus, Aquarius and Pegasus. Hardly worth the effort, but Equuleus is well placed during September.

Historically

You have to admire the astronomers of yesteryear. Perceiving the head of a foal from this motley collection of dim stars is an impressive feat of imagination. The first reference to Equuleus is in Ptolemy's *Almagest*.

Notable Stars

Double: Epsilon Equulei 5.2/7.0 and 11″. The primary is again double, but very close.

Deep Sky Objects

There are no deep sky objects of note within Equuleus.

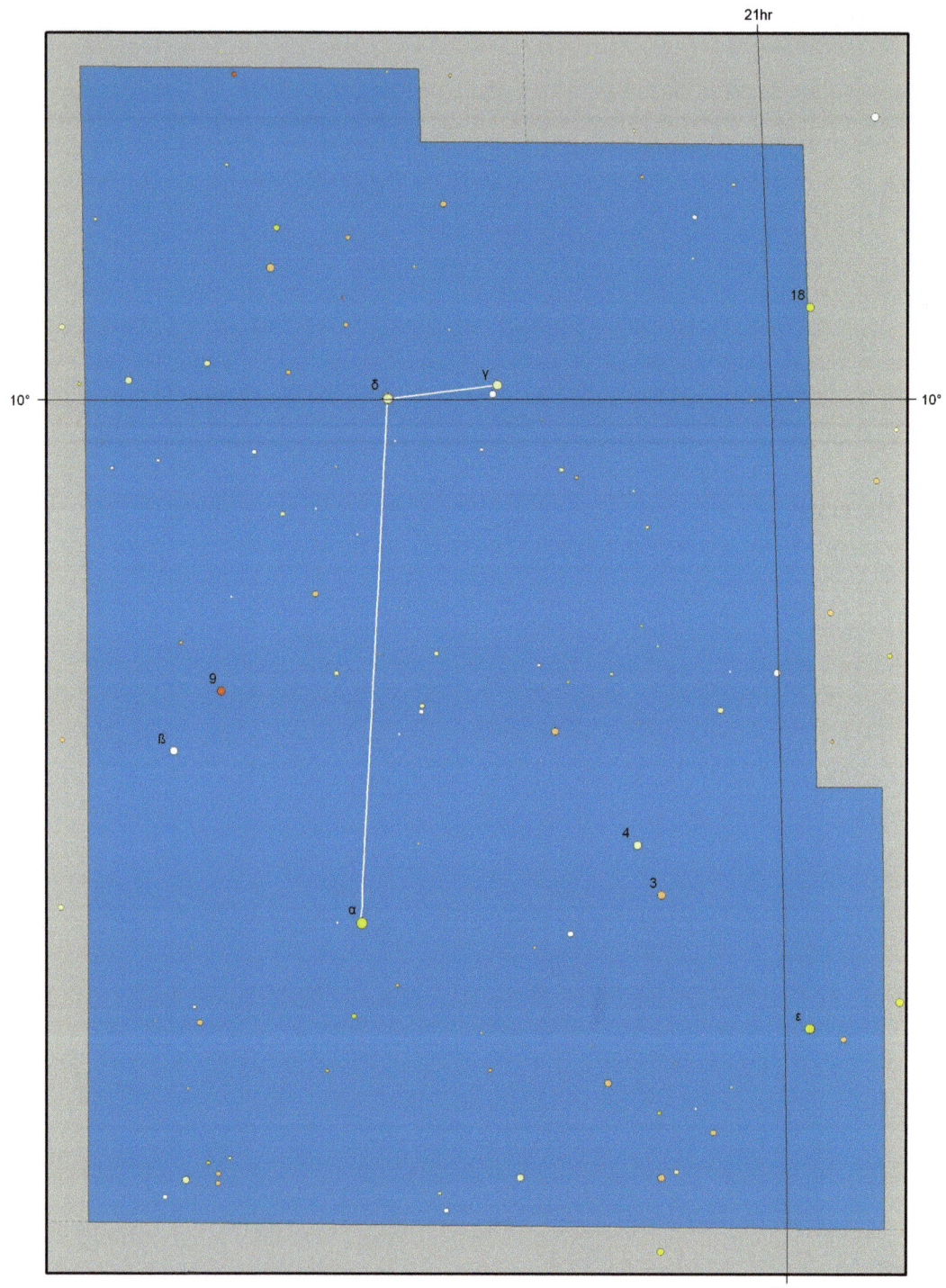

ERIDANUS

A long constellation passing from Rigel on the celestial equator, southward 60° to its end at the first magnitude star Achernar. For a constellation this large – it covers 1,138 square degrees of sky – it is relatively poor in deep sky objects suitable for smaller apertures, although nearby Orion certainly makes up for that. The constellation, as a whole, is not visible from Europe and some parts of north America.

Historically

A celestial river winding southward from Orion. It may represent The Nile or, possibly, The Po in Italy. There is a long-associated tale of Phaethon, son of Helios, driving an out of control chariot across the sky before crashing into the river.

Near Orion where Eridanus starts, there was, at one time, a constellation named Sceptrum Brandiburgicum. It was invented in 1688 by Gottfried Kirch to celebrate the province of Brandenburg and, perhaps, to get in a little deft flattery towards its ruler, Fredrick III. Kirch tried a similar trick several times and eventually ended up becoming the first director of the Berlin Observatory – appointed by Fredrick III. The constellation didn't last long, though.

Notable Stars

Double: Omicron Eridani "4.5/9.7 and 83". Fainter component further out. A triple star including one white dwarf and one red dwarf. Very nearby system.

Double: Theta Eridani 3.5/4.5 and 8″. White stars.

Nearby star: Epsilon Eridani is one of our Sun's nearest neighbors. We are just 10.5 light years away from this, which moves across the sky almost 3″ per year.

Deep Sky Objects

NGC1535: A small 9th magnitude planetary nebula 20 arc seconds across. The central star is 11th magnitude and the general form is an annulus. Some detail may be visible at high magnifications and apertures. Some report it as pale blue in color.

NGC1232: A face-on spiral in Eridanus looking much like NGC3184 and M101 in images. Tenth magnitude and large, at 10×9 arc minutes across.

NGC1291: An intriguing 9th magnitude galaxy. It appears as a dim patch 1 arc minute across with a distinct central brightening. It caused confusion for years, being called, variously, a globular cluster, an open cluster or a galaxy. It's now known to be a face-on spiral galaxy, where the outer spiral arms are experiencing a flare up of star formation, creating a ring of brightness. The bright ring is 8 arc minutes across and probably best imaged.

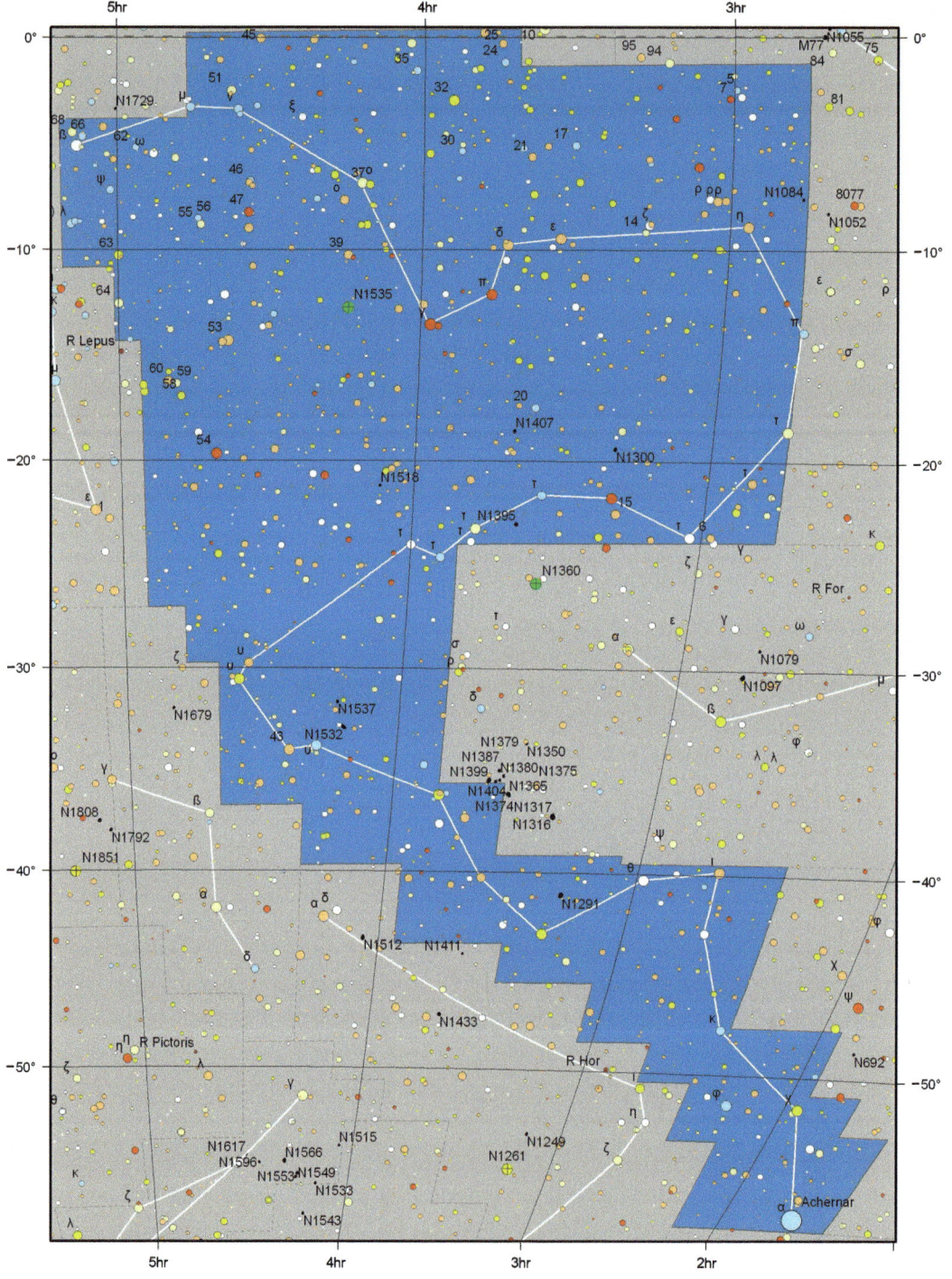

FORNAX

A southern hemisphere constellation best known for a cluster of rather dim galaxies invisible in small aperture instruments. Its brightest stars are of 4th magnitude and appear in a sparse region of sky below, and to the east of, Cetus.

Look out for it at the years' end.

Historically

Fornax portrays the laboratory furnace invented by Antonie Lavoisier. The chart he drew suggests a distillation process of some sort, which is pretty impressive given the few noticeable stars he had managed to borrow from the neighbouring constellation of Eridanus.

Between Fornax and Sculptor lie some stars of a constellation created by Johann Bode in his Uranographia atlas. It was Machina Electrica and looked rather like a Wimshurst Machine used to create static electricity.

Notable Stars

Double: Alpha Fornacis 3.9/6.9 and 4.5″. White and yellow pair.

Double: Omega Fornacis 5.0/7.8 and 11″. White and blue stars.

Sunlike star: The brighter component of Alpha Fornacis is very similar to our Sun in terms of composition and temperature.

Variable: R Fornacis is a Mira type star varying between magnitudes 7 and 13 in 389 days.

Deep Sky Objects

NGC1316: A 9th magnitude galaxy that will appear around 2 arc minutes across. It has a bright core, which may, at first, be mistaken for a star.

GEMINI

A splendid northern hemisphere constellation dominated by the pairing of its brightest stars Castor and Pollux. Embedded in the Milky Way, it's a source of pleasure for the deep sky observer, while the planetary observer will welcome the fact that the ecliptic runs through it.

During the nights bracketing the 13th December, a stream of brilliant meteors, and the occasional fireball, radiate out from a point near Castor. Well worth sitting out on a cold night for. A wonderful shower – the best of the year – and more reliable than the Perseids.

Historically

Gemini is an ancient constellation with a huge mythology; unfortunately the Greek and Roman traditions are not entirely consistent. They were twins, born to Leda after she was visited by Zeus. Depending on the version used, one – Pollux – was a mortal, as his father was Leda's husband, while Castor was immortal, his father being Zeus. It's recorded that they were inseparable crew members of the Argo and thus involved with the quest for the Golden Fleece.

Odd details abound: they may have been born from eggs, as Zeus was a swan at the time (no one said mythology had to make sense), and another pair of twins were also born, one of whom was Helen of Troy.

Notable Stars

Double: Alpha Geminorum (Castor) 2.0/2.9/9.5 and 2.2″/72″. White stars and one dim red. It's a lot more complicated than this but the other companion stars cannot be spotted with modest telescopes.

Double: Delta Geminorum 3.5/8.2 and 6.9″. Yellow and blue stars.

Double: Kappa Gem 3.6/8.2 and 7.2″. Yellow and bluish.

Deep Sky Objects

M35: A superb open cluster well worth seeing in any telescope: the bigger the better. This 5th magnitude cluster encompasses nearly half a degree of sky. It contains a number of bright stars, but the bulk of its 150+ stars are dim, providing a background glow.

NGC2158: Lying in the same field of view as M35 is another, dimmer, open cluster which is considerably further away and so appears rather more compact.

NGC2392: A nice planetary nebula, the Eskimo nebula is surprisingly easy to see despite being 10th magnitude. It appears pale blue to some observers and spans 40 arc seconds. The central star is 9th magnitude. NGC2392 should be visible in an 80mm aperture as a fuzzy edged star but is more obvious in a larger aperture.

NGC2395: An often overlooked 8th magnitude open cluster of 20 or so stars.

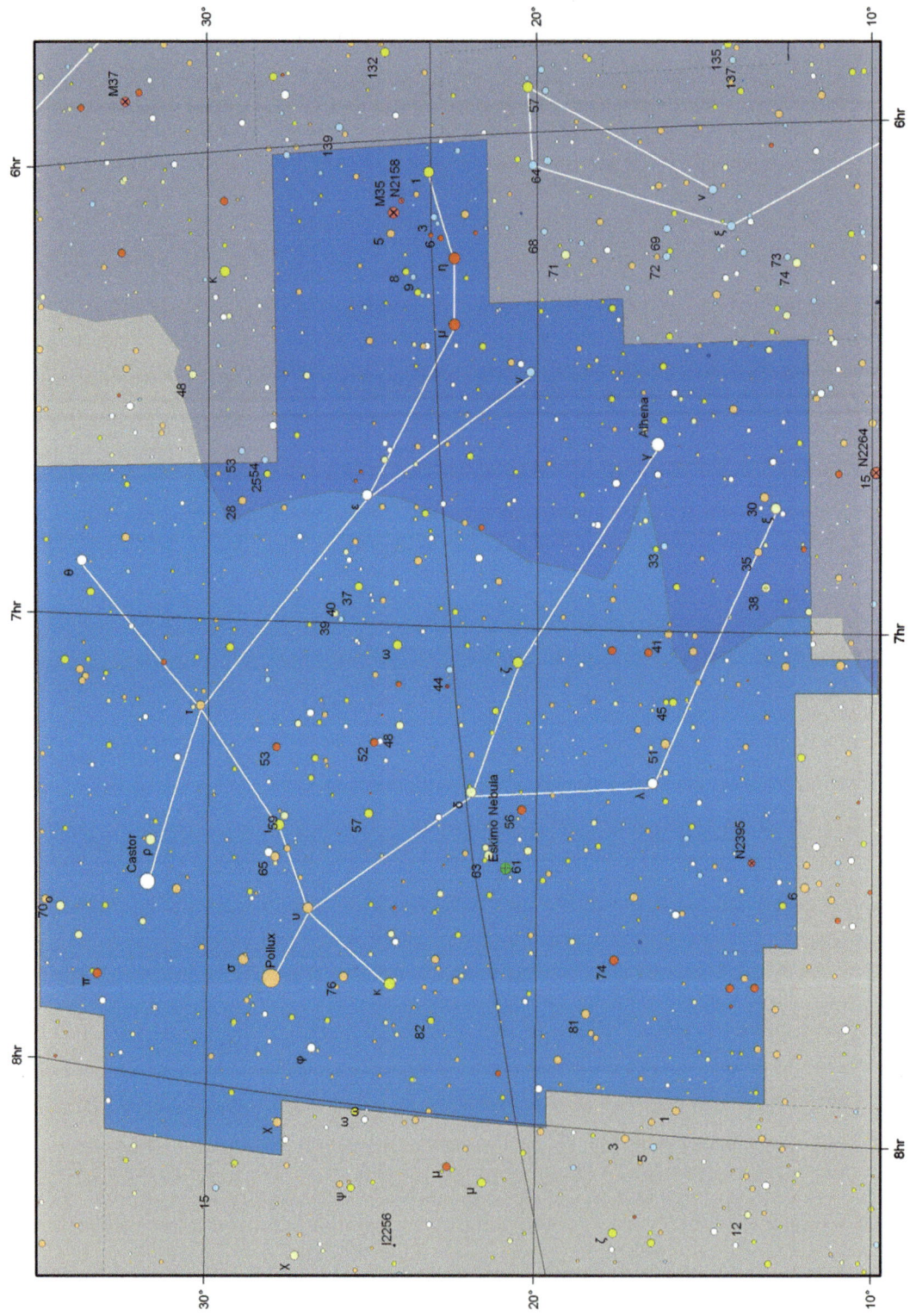

GRUS

A southern hemisphere constellation that is located just south of Piscis Austrinus and the brilliant Fomalhaut. It's bright enough to be fairly eye-catching and the triangle formed by alpha-beta-delta is surprisingly distinctive, perhaps because the sky around there is a little dull.

Best picked out in October.

Historically

Originally the stars that make up Grus, a representation of a flying bird, were part of Piscis Austrinus. Looking at the shape they create, it's not obvious why Keyser and de Houtman perceived it as a crane. Certainly it's not a patch on Cygnus. Ian Ridpath points out in his book *Startales* that they have also been represented as other birds such as a heron or flamingo.

Notable Stars

Double: Pi Gruis 6.0/5.6. One of the components is a highly variable red giant and may be brighter or dimmer than its white companion. Not a true double star but a mere alignment.

Variable: S Gruis. Another Mira type variable star with a range from magnitude 6–15 in 400 days.

Deep Sky Objects

There are no notable deep sky objects within Grus.

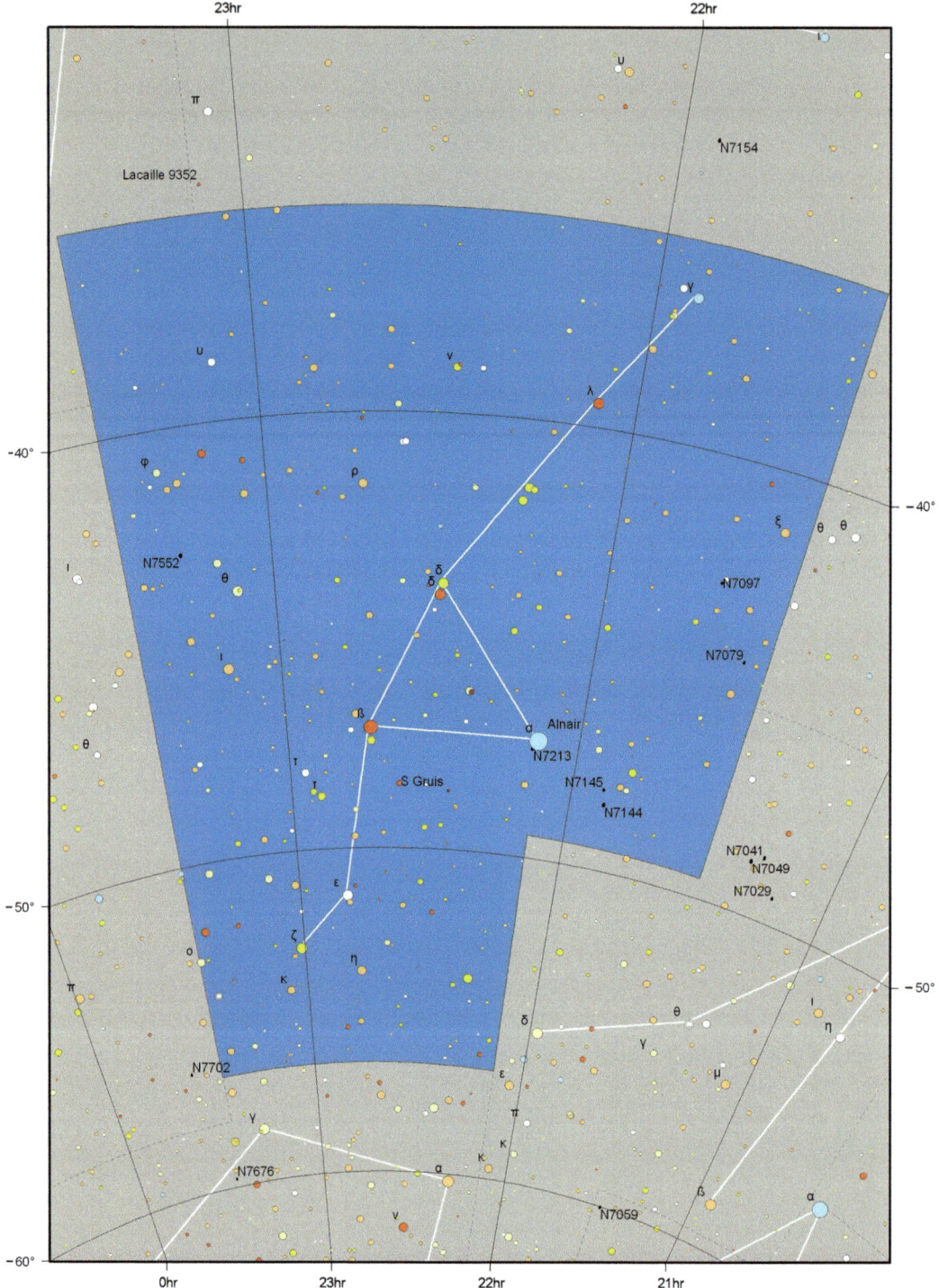

HERCULES

A northern hemisphere constellation perhaps most easily identified from the "keystone" asterism made up from pi, eta, epsilon and zeta at the heart of the constellation. None of the stars are especially bright – the brightest, beta, is magnitude 2.8 – but the shape does catch the eye. The similarly distinctive shapes of Corona Borealis and Lyra make it easy to locate.

The constellation contains much of interest for the beginner, including several fine globular clusters and a superb double star in alpha Herculis.

Historically

The constellation of Hercules (the Roman equivalent of Heracles) portrays a figure, presumably Hercules, battling a serpent and possibly forced to one knee. Actually, the constellation predates that Grecian view. The oldest-known records refer to it as a kneeling figure. It's unclear whether the figure was entirely human, initially, but it has evolved to be a man over the centuries.

Hercules is another child of Zeus – in this case, Zeus disguised himself as someone else to fool the victim of his attentions, who was his own great granddaughter.

Notable Stars

Double: Kappa Herculis 5.3/6.5 and 28″. Yellow stars.

Double: Delta Herculis 3.1/8.2 and 8.9″. White and blue.

Double: Rho Herculis 4.6/5.6 and 4.1″.

Double: Alpha Her 3.5/5.4 and 4.7″. Red/Bluish. Beautiful contrast.

Double: Mu Herculis 3.4/10.1 34″. Yellow and orange.

Deep Sky Objects

M13: The most impressive northern hemisphere globular cluster. It is visible as a hazy naked eye 5th magnitude star on the edge of the Keystone. Obvious in a small telescope with dozens of 10–11th magnitude stars to be seen. A lovely sight in larger telescopes as it spans a third of a degree. Not as spectacular as Omega Centauri, but superb nonetheless.

M92: A fine and compact 6th magnitude globular cluster. Really worth looking at, as it resolves readily with a 150mm aperture. For an object 27,000 light years away it's a wonderful sight.

NGC6210: A planetary nebula a mere 10 arc seconds across and 8th magnitude. Visually little detail will be present but it will be easy to spot.

NGC6229: Another globular galaxy. Shining at 9th magnitude and appearing just 2 arc minutes across, a 300mm aperture will be needed to resolve the individual stars, which are 15th magnitude.

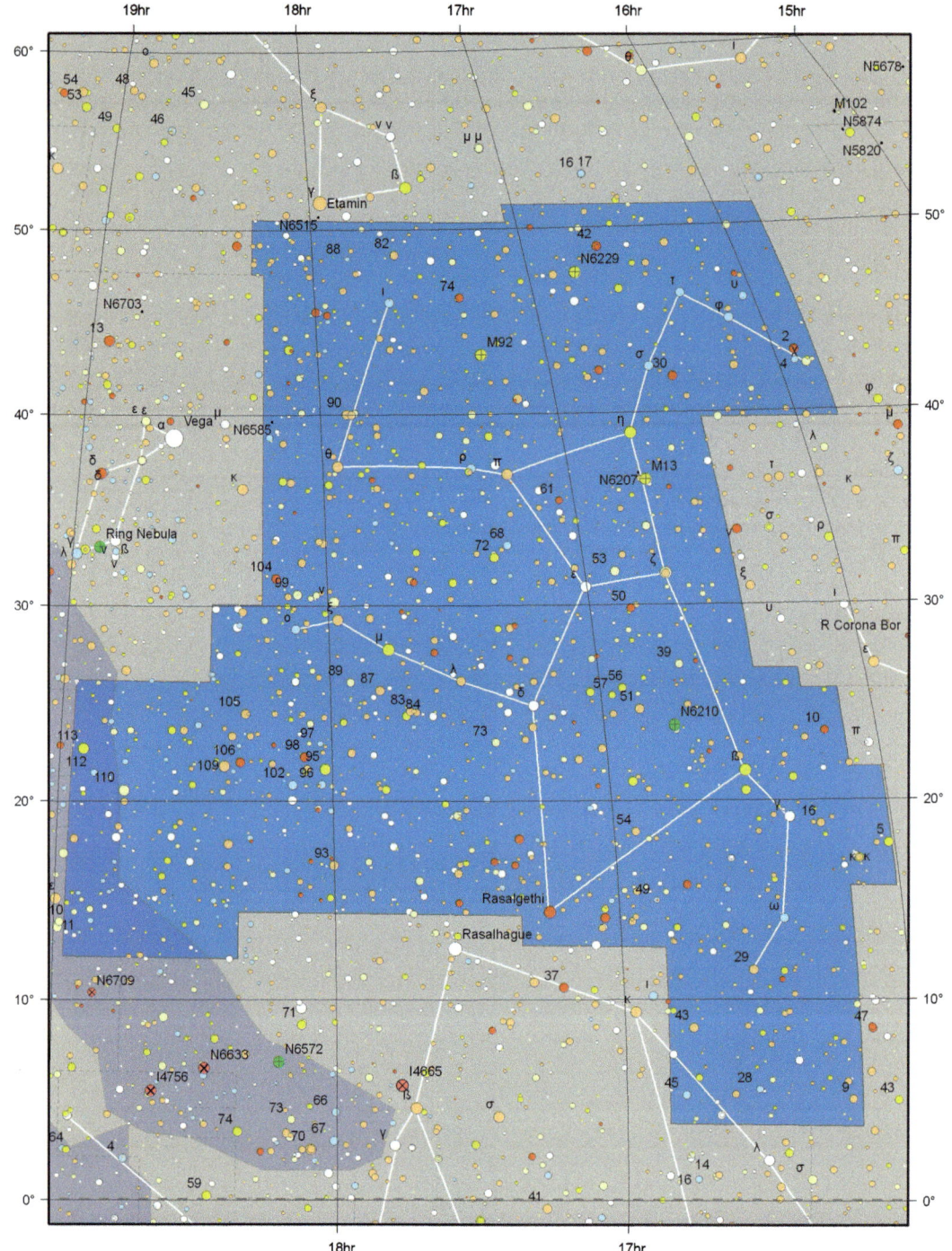

Fig. 14 The globular cluster, M13 in Hercules was known to the Ancient Greeks and can be seen in the "Keystone" of the constellation (Image by Grant Privett)

HOROLOGIUM

A southern hemisphere constellation. Alas, it isn't very bright and doesn't have an obvious form to it.

Historically

Horologium originally rejoiced in the name Horologium Oscillitorium and commemorated the invention of the pendulum clock by Christian Huyghens – who is better known in astronomical circles for discovering Saturn's rings. Fortunately, over the years, part of the name has been dropped.

The constellation was created by de Lacaille who, were he mapping the sky today, would probably be producing constellations named the Laptop, the Ipad and the DSLR. He *really* liked the technology of the time.

Notable Stars

No double stars.

Variable: R Horologii is a red giant star and a Mira type variable. Its brightness range, magnitude 4–14, is one of the largest known.

Deep Sky Objects

NGC1261: The only deep sky object in Horologium is a 9th magnitude globular cluster discovered by James Dunlop in New South Wales, Australia, during 1827. Visually, it is nearly 2 arc minutes across, with a 9th magnitude yellow star just 4 arc minutes away. It has a compact core.

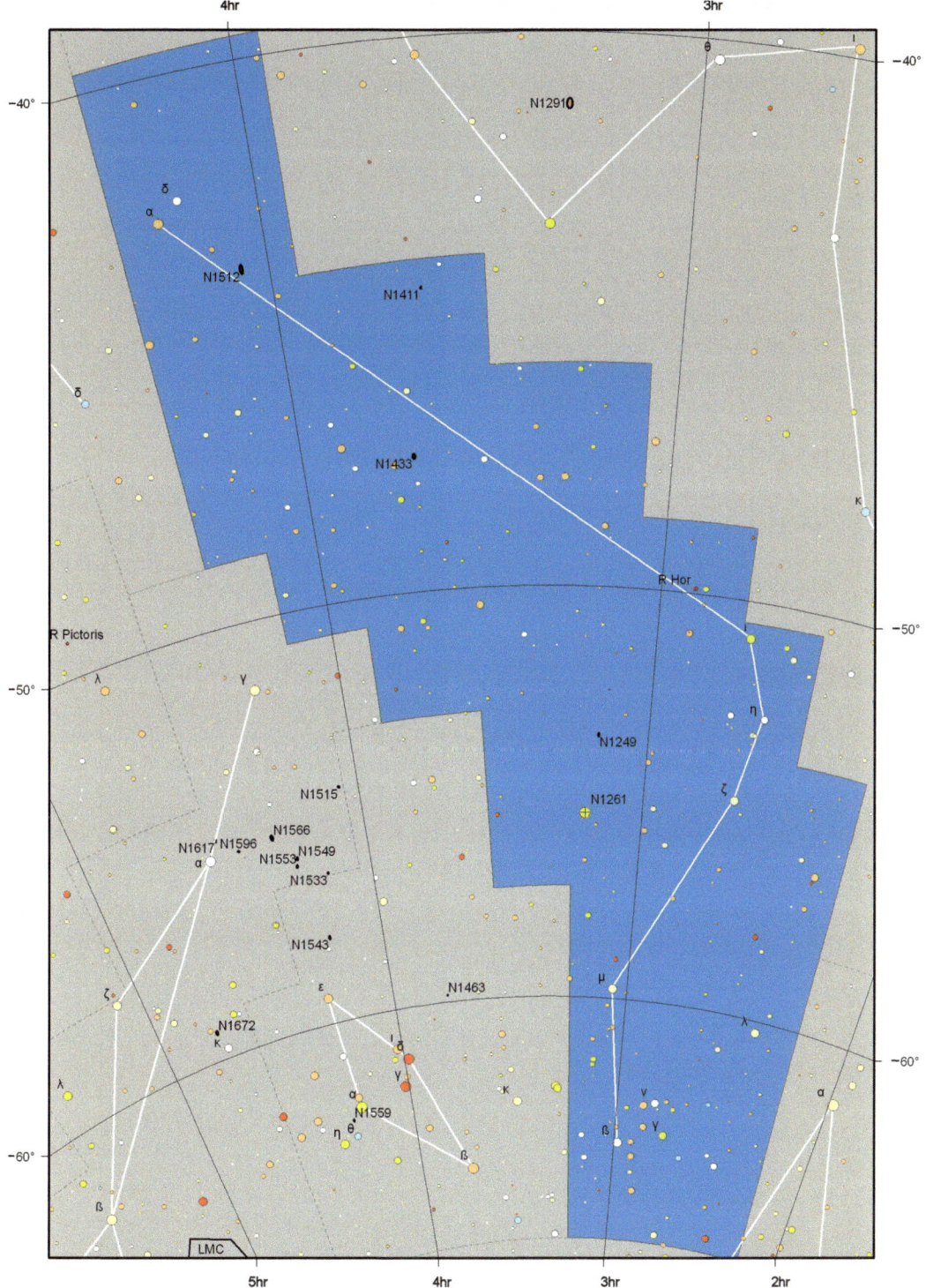

HYDRA

A huge sprawling run of stars crossing 7 hours of right ascension. The constellation is the largest in the sky and has several deep sky objects to keep you interested. The main stars of this predominantly southern constellation are 2nd, 3rd or 4th magnitude. The head of the serpent crosses into the northern hemisphere. From the UK its serpentine form runs nicely along the southern horizon during the months March through May.

Historically

An ancient constellation representing a serpent. Which particular serpent depends upon the culture you consider. The hero tasked with taking on the monster is generally assumed to be Apollo or Hercules for the Greeks or Romans, but the name of the hero involved in any equivalent Babylonian legend is now lost in time.

Notable Stars

Double: Theta Hydrae 4/9.9 and 29″

Deep Sky Objects

M83: A splendid face-on barred spiral 7th magnitude galaxy lying six times further away than the Andromeda galaxy, at 15,000,000 light years distance. It is visible in binoculars and spans 8 arc minutes. One of the brightest galaxies in the sky. A must see object

NGC3242: A ring-like planetary, looking a little like a dim M57. The central star is a surprisingly bright 11th magnitude and the annular ring approximately 40 arc seconds across. Very easily located close to Mu Hydrae.

M48: A bright open cluster that borders on naked eye visibility. Its 30 arc minute width contains many stars of 8th magnitude or fainter. Thirty or so member stars will be seen in small instruments. Obvious in binoculars.

NGC5694: A 10th magnitude globular cluster spanning 2 arc minutes. The individual members are too faint for all but the largest telescope.

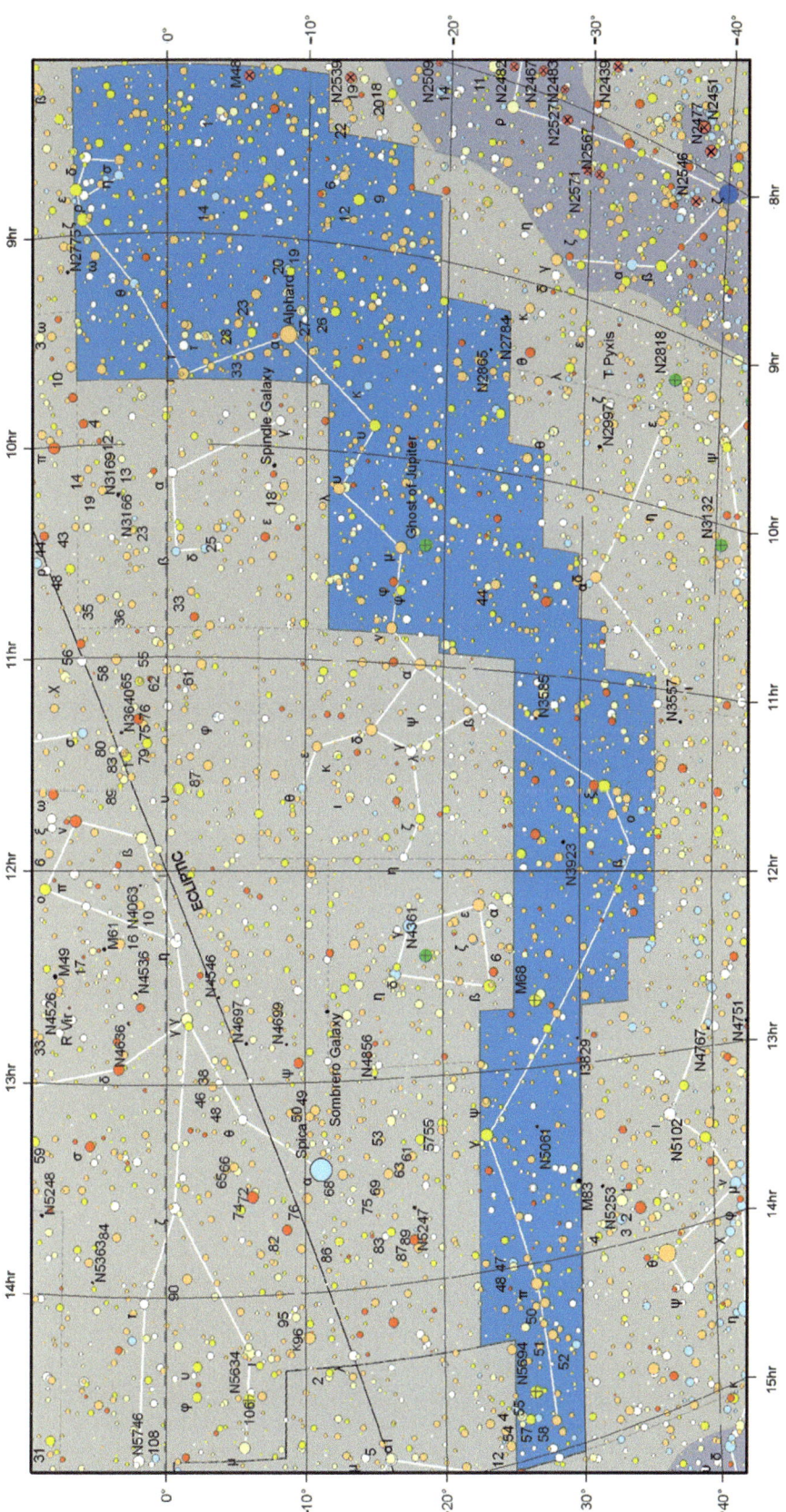

HYDRUS

A southern hemisphere constellation, nestling between Achernar at the southern extreme of Eridanus and the South Celestial Pole in Octans. Its brighter 2nd and 3rd magnitude stars form a quite distinctive shape. Besides a binocular double star, Pi, the constellation is not well provided with other distractions.

Historically

Hydrus represents a small snake. It nestles between the Magellanic clouds and is little more than a scattering of stars and a triangle. It was introduced by Plancius, who depicted it on a globe in 1598.

Notable Stars

Double: Pi Hydri. 5.6/5.8 Yellow and orange stars. Not a true binary star but an alignment of stars at very different distances from us.

Deep Sky Objects

NGC602: A small open cluster that can be detected in 80mm binoculars. It appears to lie within the outer reaches of the Small Magellanic Cloud, which dominates neighboring Tucana. There is an associated nebula – 30 arc seconds wide – which is more likely to be spotted than the cluster.

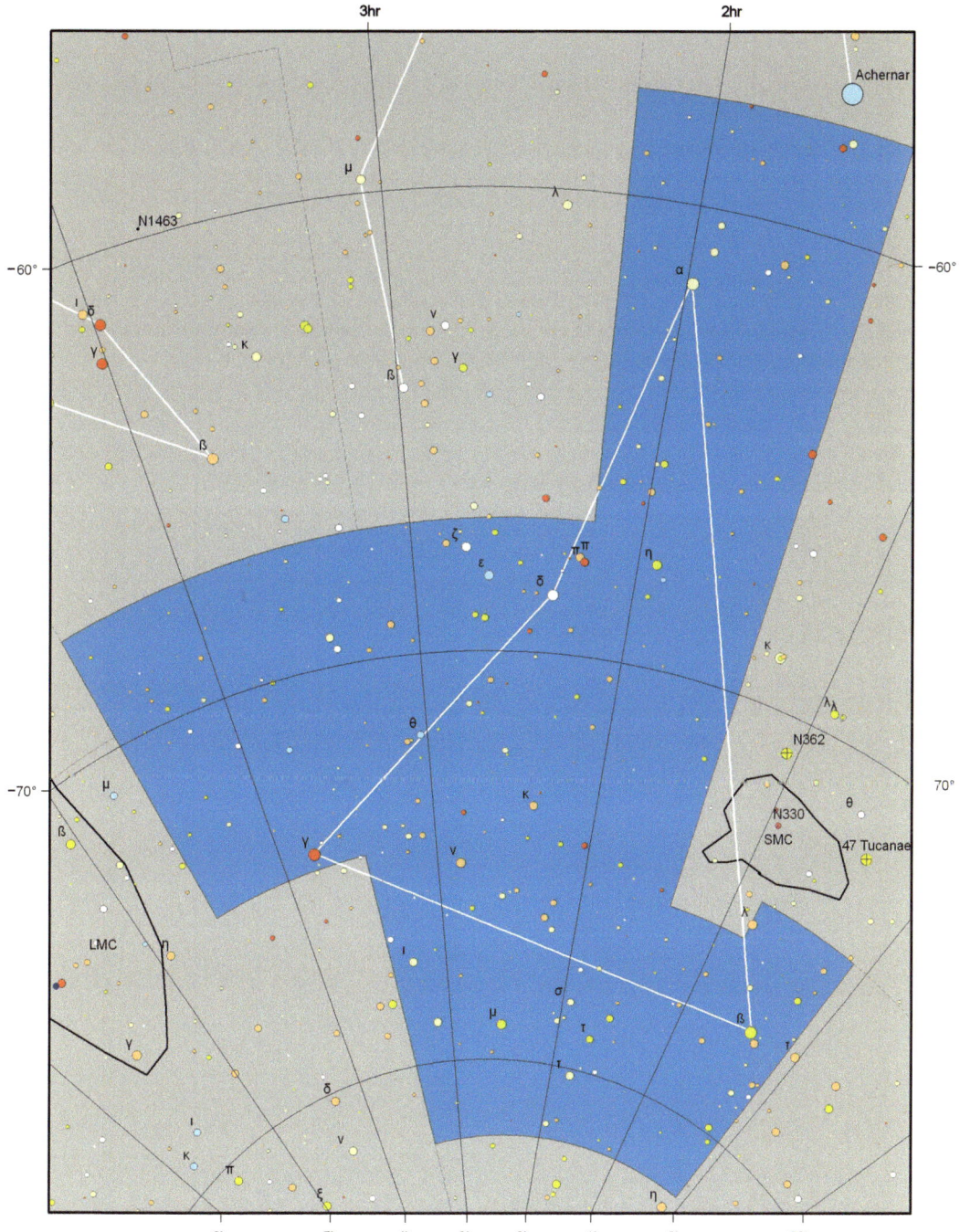

INDUS

A southern hemisphere constellation to be found just west of the more eye-catching form of Grus. Its major stars are 3rd or 4th magnitude. The brightest star in Pavo, a constellation to its west, makes locating the constellation rather easier. Look out for it in September.

Historically

Yet another constellation arising from the globes and maps created by Petrus Plancius, based upon observations by Pieter Keyser and Frederick de Houtman. It represents a native of one of the lands visited, holding a spear. It's one of the less well known male figures in the sky.

It's worth remembering that the voyage Keyser and de Houtman undertook was, in part, motivated by setting up vastly profitable trades between Europe and a source of spices, i.e.: Indonesia. It certainly wasn't a walk in the park or a scientific expedition, as the brutal story of Frederick de Houtman's brother, Cornelis, testifies. It is worthy of a book, or film, in its own right.

Notable Stars

Double: Theta Indi 4.5/7.0 and 6.5″. Yellow and red.

Nearby star: Epsilon Indi is the 17th nearest star to Earth. It lies just 12 light years away and is orbited by two failed stars called brown dwarfs. It is also moving quickly across the sky. In just six centuries time it will enter the constellation Tucana – that is pretty quick by astronomical standards.

Deep Sky Objects

There are no notable deep sky objects in Indus.

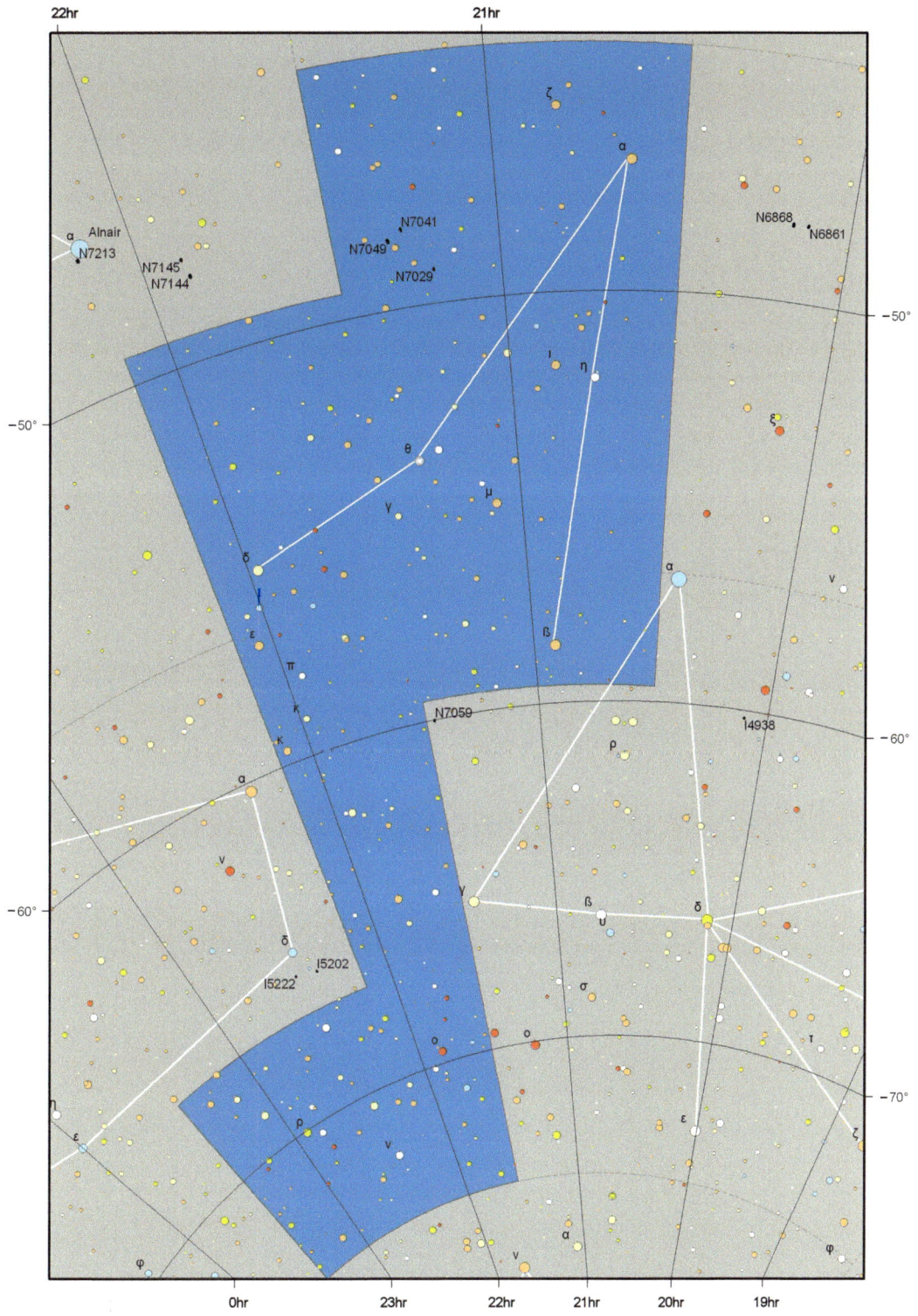

LACERTA

A northern hemisphere constellation formed of a zig-zag pattern of stars running north to south across almost 20° of sky. It occupies the space north of Pegasus, between Cygnus and Andromeda, and so is circumpolar from much of Europe and north America. Beyond that, this collection of 3rd, 4th and 5th magnitude stars doesn't have a lot going for it.

Best located for viewing during the month of October.

Historically

Lacerta was put together by Johannes Hevelius in 1687 and appeared in the star atlas Firmamentum Sobiescianum. It represents a lizard, but there are no particular references to one in mythology, so it is not clear why. However, unlike many of his constellations, this one was adopted and is still used today.

Hevelius has the distinction of being the first Polish citizen to be elected to the British Royal Society.

Notable Stars

No double stars.

Deep Sky Objects

NGC7243: A reasonably bright and young open cluster. It is easily visible in binoculars and appears to be 20 arc minutes across. At the heart of the cluster is a variable star accessible to a 150mm aperture. A clumpy cluster of more than 25 members.

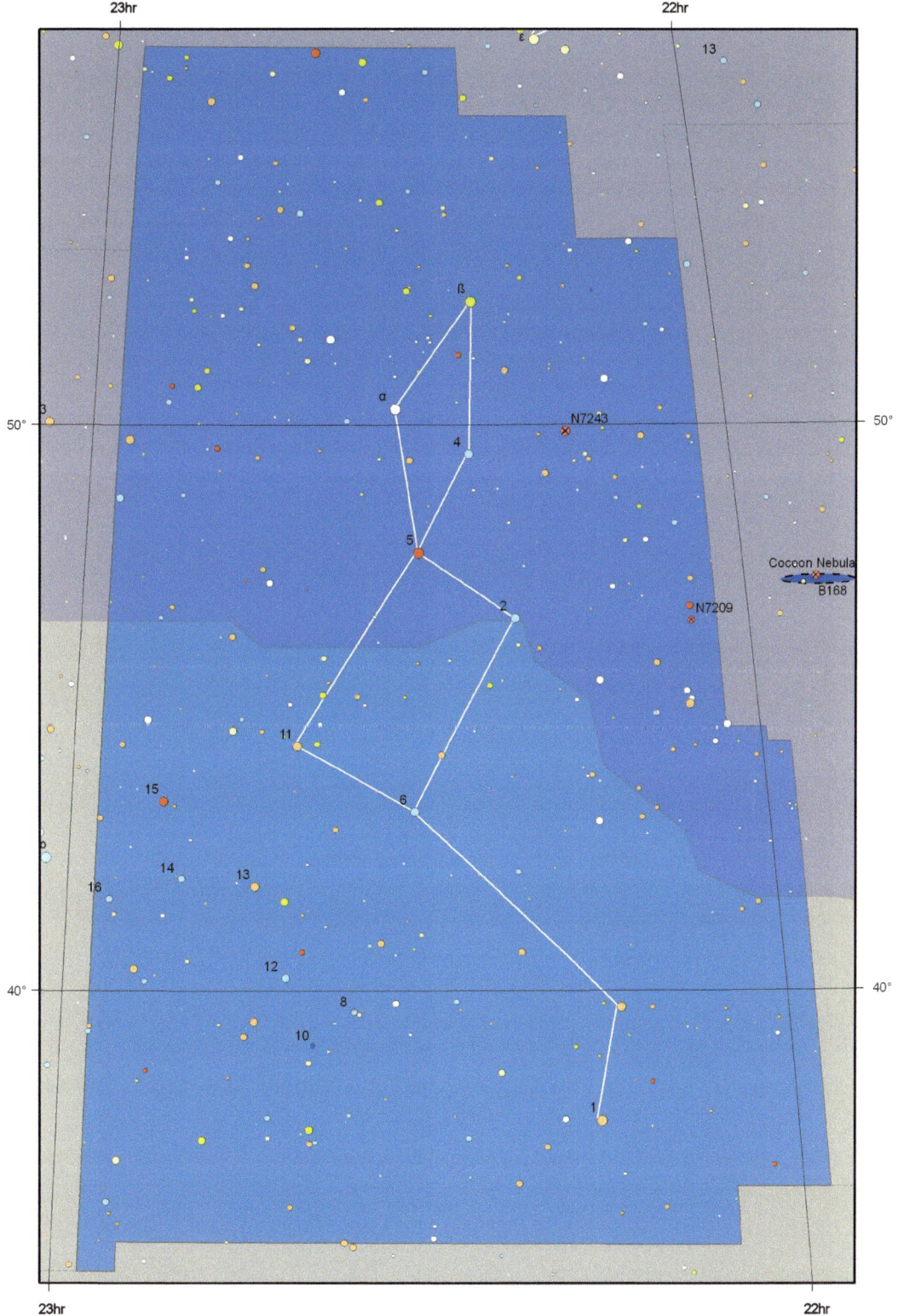

LEO

One of the classic northern hemisphere constellations. Its distinctive and bright form is dominated by the 1st magnitude star Regulus, the 2nd magnitude Denebola and the "sickle" shape formed by the stars running from Regulus through to lambda Leonis. A lovely sight on April evenings.

Every 33 years a massive outburst of meteors appears in mid November – thousands of meteors an hour may fall as debris from the Comet Tempel/Tuttle as they hits the atmosphere. The rest of the century is less impressive, but there is positive activity from a radiant within the sickle in lesser years. Look for the display on the 17th November. The *Handbook of the British Astronomical Association* can give guidance on the year's prospects.

Historically

Leo is an anciently-described region. It has consistently been deemed to portray a Lion. Inevitably, perhaps, it has been seen as the lion killed by Hercules in the course of his Labors, but each culture would have had its own myth:- people of all cultures love a good yarn! As lions are perceived as the king of the beasts, its presence in the sky is not unexpected.

Notable Stars

Double: Alpha Leonis (Regulus) 1.4/7.9 and 177″. Orange

Double: Gamma Leonis 2.4/3.5 and 4″. Yellow stars.

Deep Sky

M66: The brightest of three closely spaced bright galaxies the others being M65 and NGC3628. All three galaxies, sometimes called the Leo Triplet, can be seen in the same low power field of view. M66 appears elliptical at the eyepiece and is in fact a spiral. Look for an 8th magnitude fuzzy patch 3×1 arc minutes across in smaller instruments.

NGC2903: Another inclined spiral galaxy. At 9th magnitude, it's surprising Messier didn't see it. It will appear elongated and, perhaps, 6 arc minutes in length, with a bright compact core. With 200-300mm apertures, detail will be detected as "mottling" within the galaxy. The dark lanes meant that Herschel thought it was two distinct objects and catalogued both. Visible in binoculars…just.

M96: A 10th magnitude galaxy that will require an aperture of 100mm or more to make its core visible. Not as obvious as NGC2903 or M66. The dimmer M95 face-on barred spiral lies nearby.

M105: An elliptical galaxy of 10th magnitude not far north east of M96. It will appear similar to an unresolved globular cluster 2 arc minutes wide.

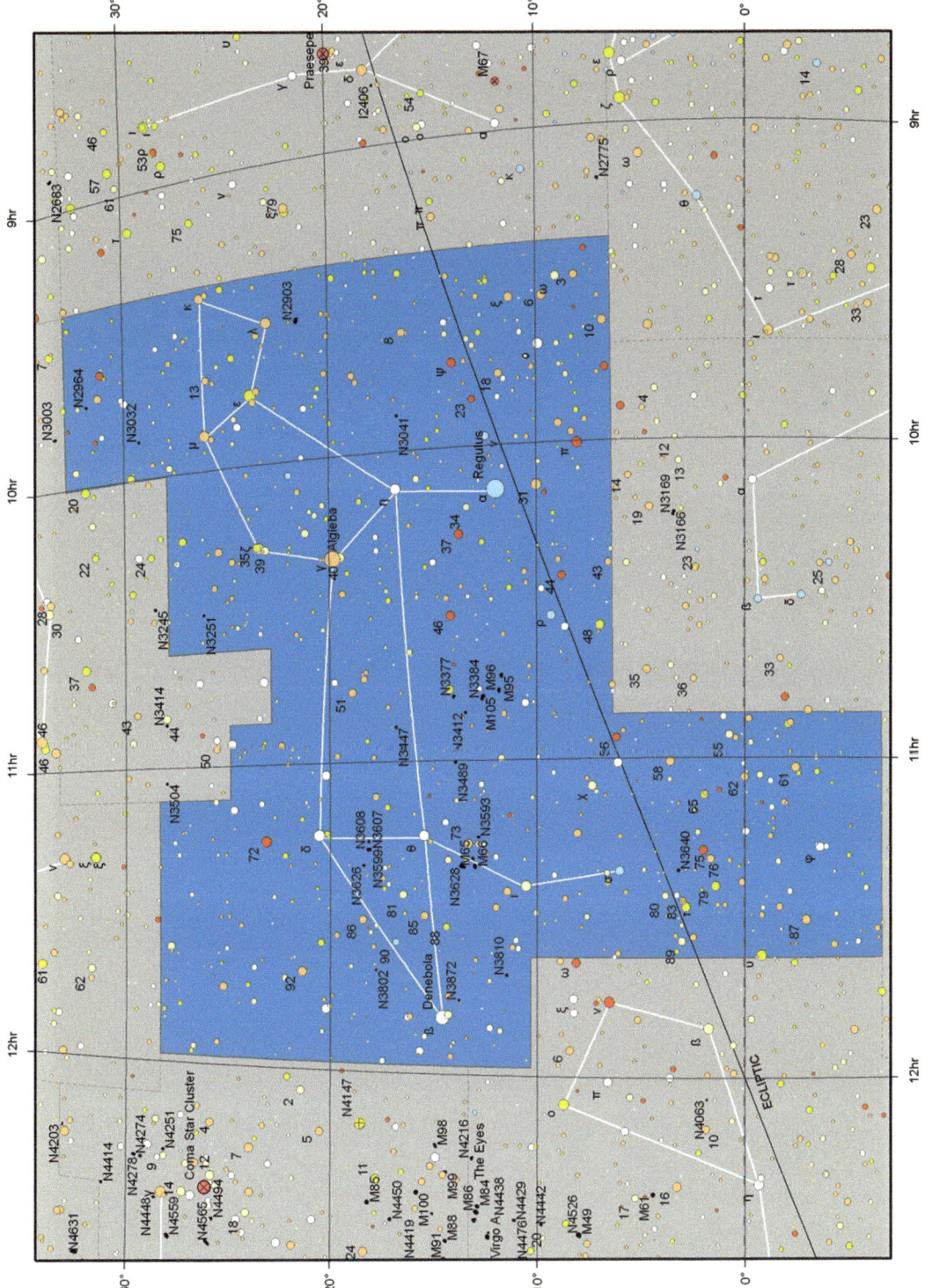

Fig. 15 M65 and M66 lie in the same low power field of view within the constellation of Leo (Image by Bill Snyder)

Fig. 16 An inclined spiral galaxy, NGC2903 in Leo boasts star forming regions visible in quite small telescopes (Courtesy Grant Privett)

LEO MINOR

A small unprepossessing northern hemisphere constellation that rather lacks a clear outline and has little in the way of other features or attractions. It's best seen during the month of April but, frankly, it's hardly worth the effort for smaller apertures.

Historically

A rather less impressive collection of stars than its larger counterpart. Leo Minor was an invention of Hevelius, who scraped it together from a collection of stars of 4th magnitude, or fainter. The shape doesn't look much like a lion, but it could have been worse. According to the author Ian Ridpath, this area was once visualized as a eunuch by the ancient Chinese map makers.

Notable Stars

No double stars.

Variable: R Leo Minoris varies between magnitudes 6 and 13 in about 370 days. A Mira type star – it's a red giant.

Deep Sky

There are no worthwhile deep sky objects in Leo Minor.

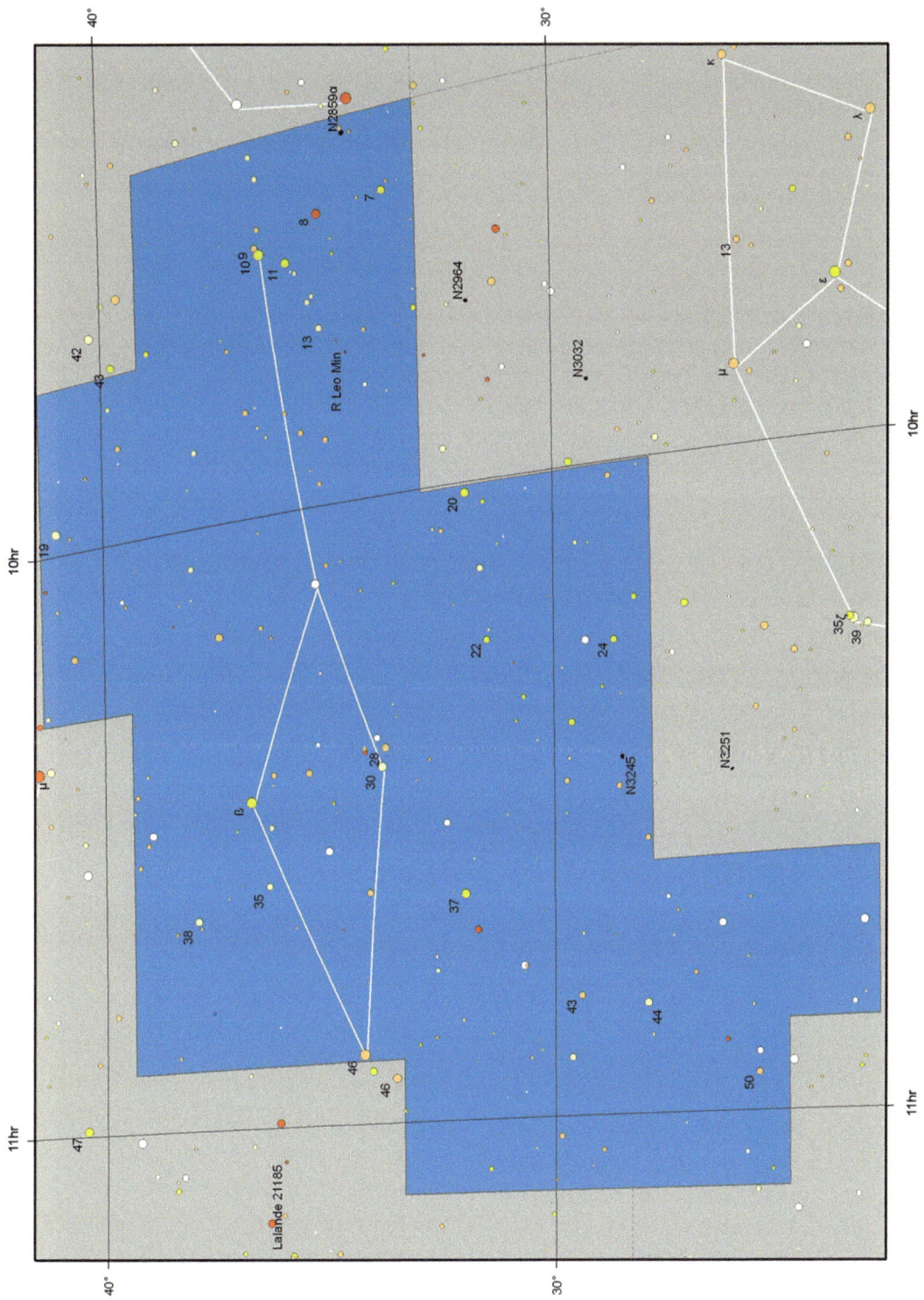

LEPUS

A southern hemisphere constellation nestling below the feet of Orion and west of Canis Major. It's not a large constellation and is, inevitably, overlooked somewhat in such grand company, although its two 2nd magnitude stars make it easy to trace out. A couple of decent double stars, and some nice deep sky objects, make it worth visiting.

Historically

The distinctive form of Lepus was described by Ptolemy in his book, the *Almagest*, over 1,800 years ago. While this identification has been accepted for a long time, there doesn't appear to be much in the way of legend associated. It's not unreasonable to assume, however, that a hunter such as Orion (adjacent) would have some of his quarry close to hand.

Notable Stars

Double: Gamma Leporis 3.6/6.3 and 95″. Yellow and orange stars.

Double: Kappa Leporis 4.5/7.3 and 2.6″. White and bluish.

Variable: R Lepus is a Mira type star sometimes known as Hind's Crimson Star. It varies between magnitudes 5 and 11 over a period of roughly 430 days. It becomes reddest when dim.

Deep Sky Objects

M79: The only obvious deep sky object in Lepus is a fine globular cluster of 8th magnitude. Notable for the high concentration of stars near its centre. Spectroscopic studies suggest it may be in orbit about one of the Milky Way's satellite galaxies. Visible in binoculars or a 50mm finder.

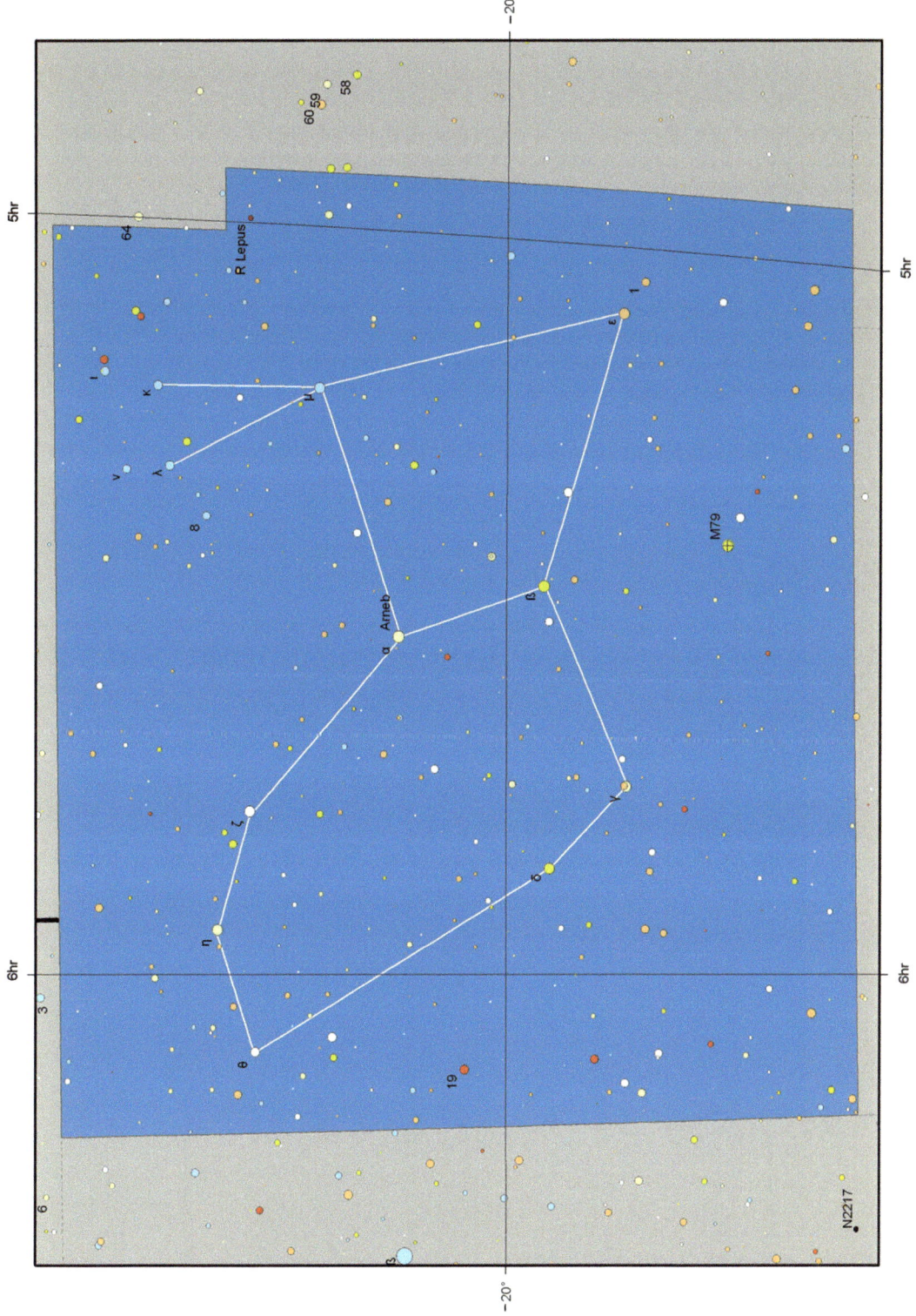

LIBRA

One of the 12 ancient zodiacal constellations, so it's no surprise the ecliptic passes through the heart of the constellation. Unfortunately, It's not seen against a Milky Way background but, with a brightest star (the magnificently named Zubenelgenubi) of magnitude 2.8, it's fairly easy to see. Other stars include Zubeneschamali and Zubenelakrab – all the names relate to claws – and hark back to its' Greek origins.

Most easily located during the nights in the middle of the year.

Historically

Libra is a constellation that was created by grabbing part of the Grecian Scorpius: specifically, the distinctive claws; and viewing them as a separate constellation. It appears to have occurred when Rome was great. Libra does look like part of a scorpion: why fabricate some Scales? Oddly, the Babylonians also had a constellation akin to Libra. Why remains unclear, sadly, as scorpions were probably a familiar creature!

Notable Stars

Double: Alpha Librae 2.9/5.2 and 231″. White and pale yellow stars. Some claim this star is pale green in color but this probably isn't truly the case. The vast majority of observers perceive it as a delicate blue.

Deep Sky Objects

NGC5897: A 9th magnitude globular cluster lying over 40,000 light years from us it, nonetheless, appears quite large, 6–10 arc minutes depending on aperture. A 250mm may be required to resolve its 14th magnitude brighter members.

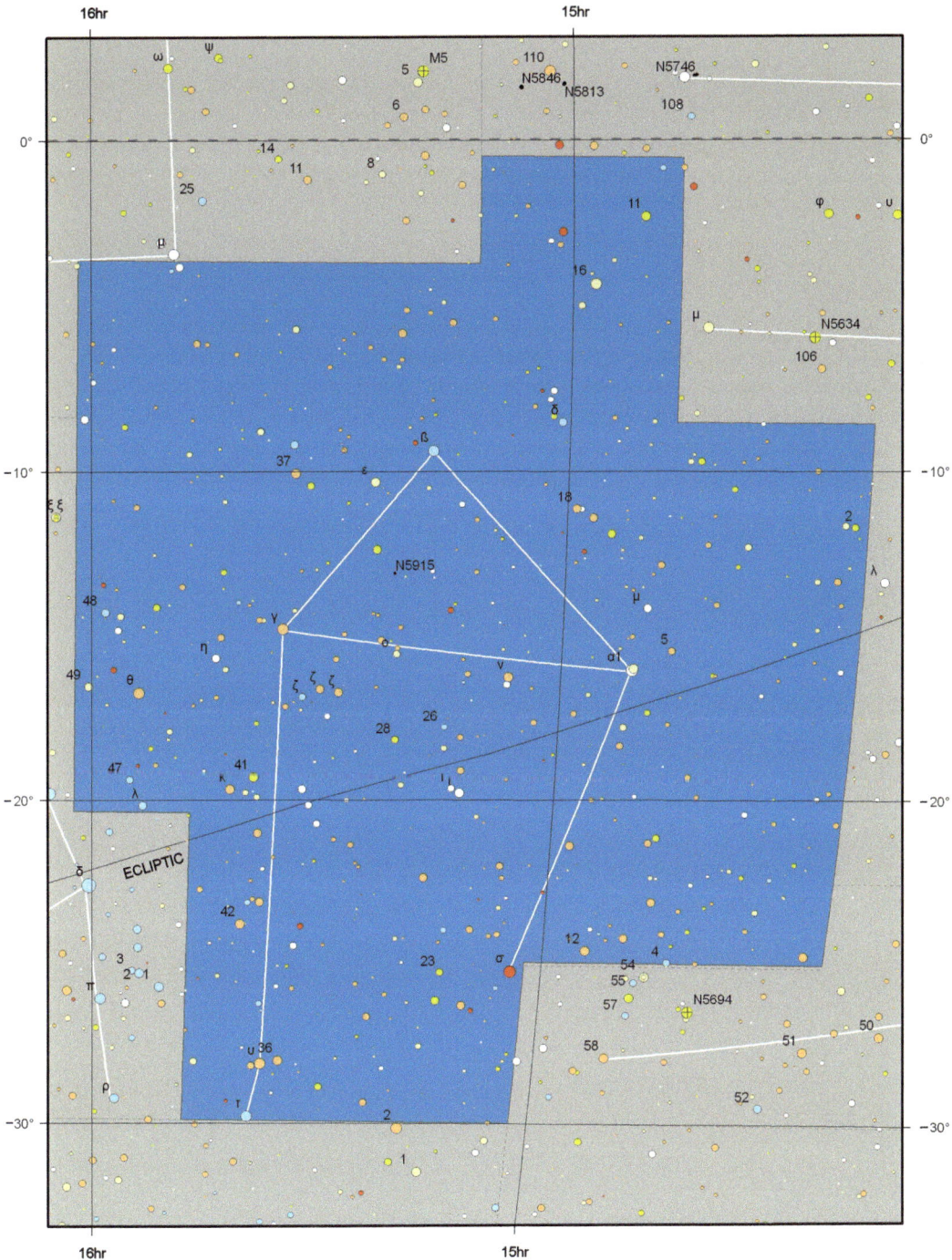

LUPUS

A southern hemisphere constellation laying south of Libra and, in part, north of alpha Centauri. Its stars are quite bright and, at 2nd magnitude, stand out well in a region of sky that is perhaps the most impressive in the sky. The Milky Way runs through the constellation.

For a fairly small constellation it has quite a lot going for it. Look out for it during the month of June.

Historically

The constellation of Lupus, the wolf, has represented a wild animal as long as records exist. When Ptolemy wrote his star catalogue 1,800 years ago he included Lupus. Prior to that, it was often seen merely as a wild animal being hunted. Inevitably, the creature depicted by the Babylonian constellation had a fairly weird combination of characteristics: it was assembled as a sort of pick 'n mix monster. A head of this; a tail of that. The stuff of nightmares.

In many accounts, the creature was impaled on a spear.

Notable Stars

Double: Kappa Lupi 4.0/6.0 and 27″. Bluish component stars.

Double: Xi Lupi 5.3/5.8 and 10.4″.

Double: Eta Lupi 3.4/7.8 and 15″. Blue and yellow stars.

Deep Sky Objects

NGC5927: An 8th magnitude globular cluster, whose brightest member is 15th magnitude. It's seen against a rich background field. Discovered by James Dunlop in 1826. May appear 6 arc minutes across in images, but perhaps only 2 arc minutes, visually.

NGC5986: 7th magnitude globular cluster. Just visible in 10×50 binoculars. A 150mm aperture should start to resolve it.

NGC5822: A large, rich and bright cluster. Sixth magnitude and over 5 arc minutes across. Brightest stars: 9th magnitude. Well worth a visit and splendid in a 150mm aperture. Easily spotted in binoculars.

NGC5823: A nice 7th magnitude cluster some 10 arc minutes across.

NGC5925: An open cluster some 20 arc minutes across. A loose grouping, where the cluster appears almost fragmented. Eighth magnitude, it contains 60 or more stars.

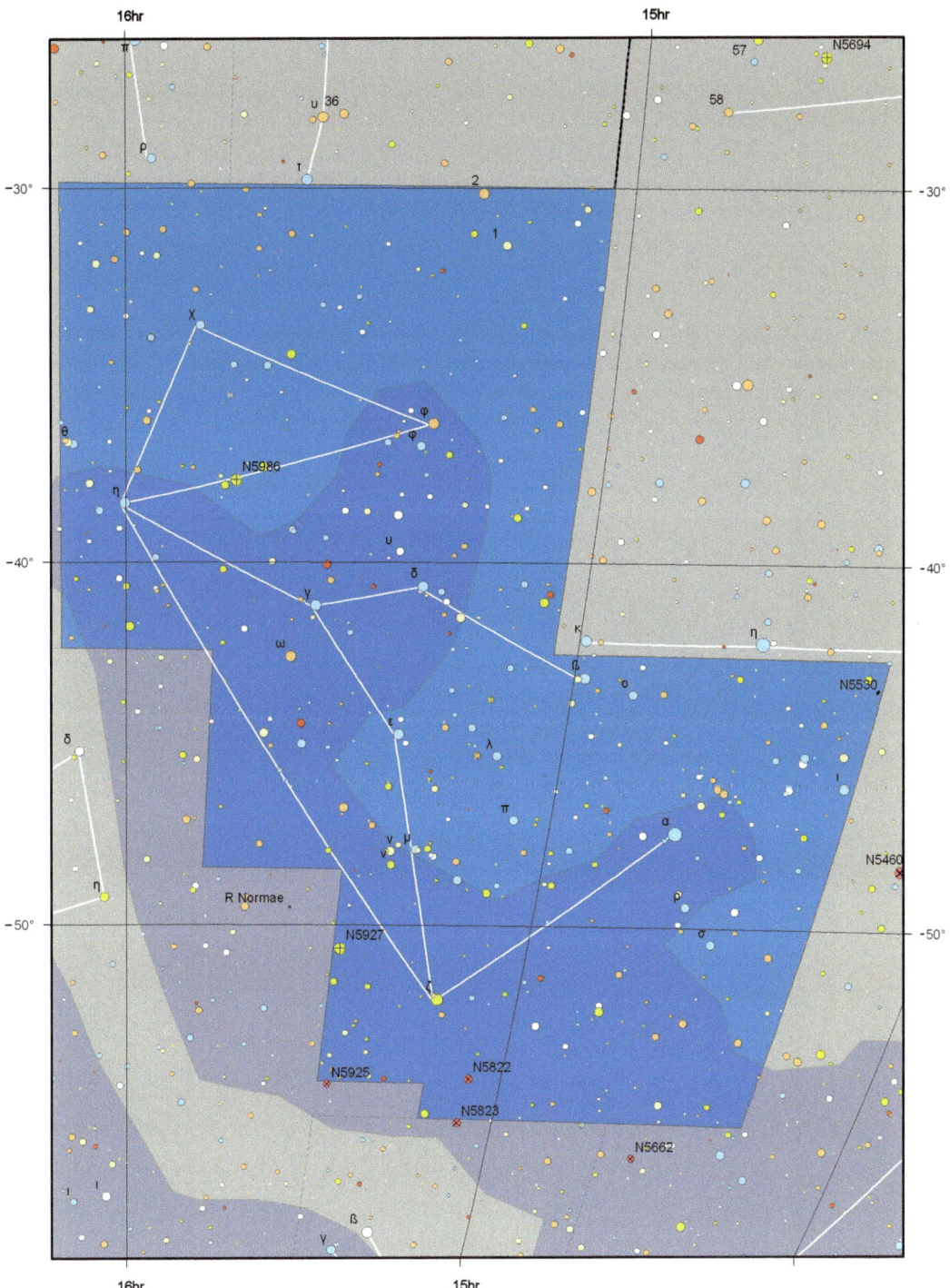

LYNX

A constellation of the northern hemisphere formed from a wiggling line of stars (the brightest 3rd magnitude) lying between Ursa Major and Gemini, and Auriga. It's not impressive, and far from memorable.

However, look for it to be at its best during the month of March.

Historically

A creation of the Polish astronomer Hevelius. It is a fairly dim and undistinguished constellation and, as a modern creation, has no legends or mythology associated with it. It is often said that to have the eyes of a lynx is to have superb eyesight. It's only true up to a point: one lynx species has very average eyesight!

Notable Stars

No double stars.

Deep Sky Objects

NGC2419: One of the dimmest objects in this book, appearing only because it's also one of our galaxy's most remote globular clusters. Appears at 10th magnitude and small – visually, but 1 arc minute across. The brightest member is far too dim for most telescopes – not surprising, with the cluster located some 300,000 light years away.

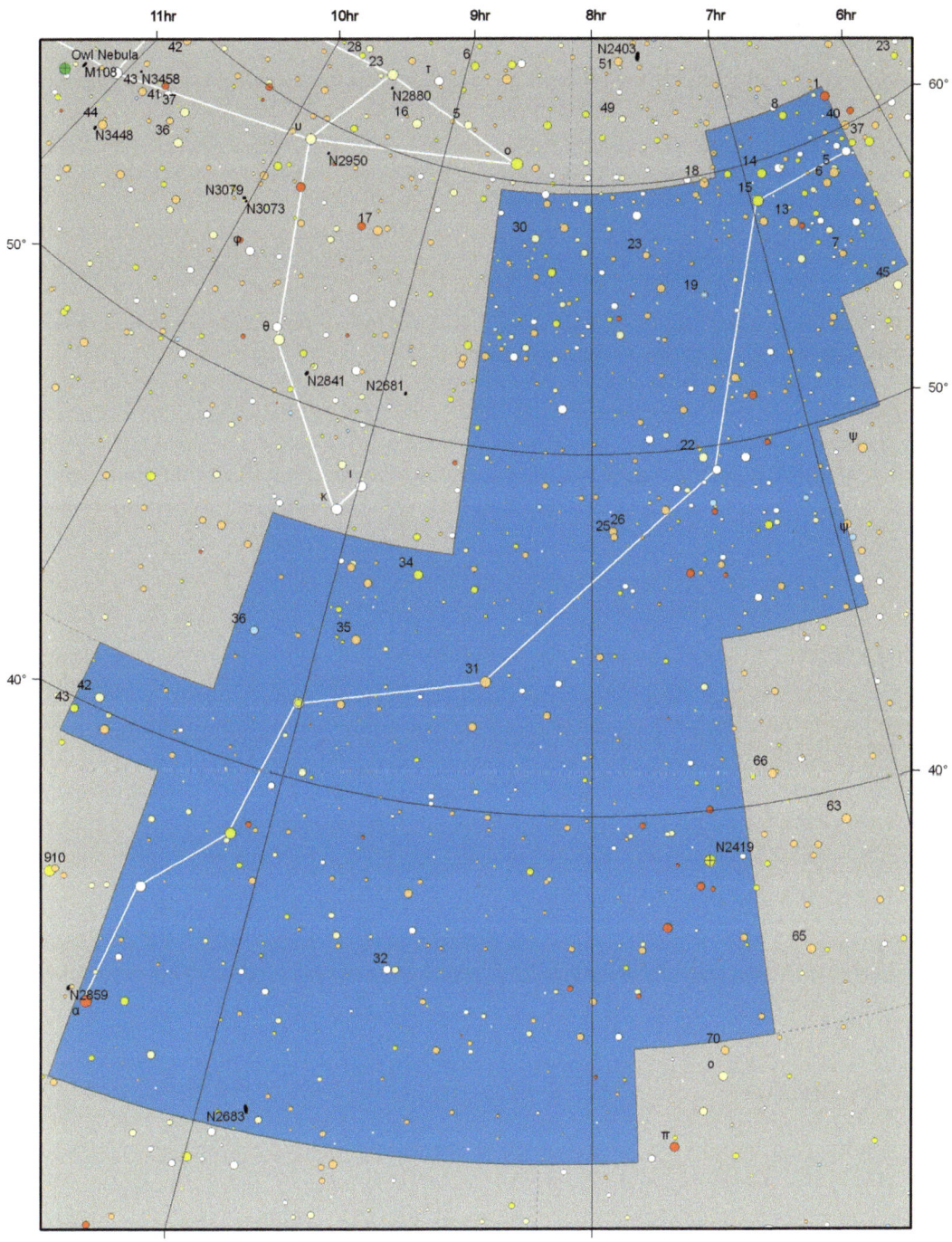

LYRA

A northern hemisphere constellation. Its lead star, the blue Vega, is circumpolar from northern parts of Europe and the USA. Vega is also a brightness reference for the whole sky – hence its magnitude of 0.0. The shape of Lyra is quite distinctive and the rhomboid of stars very obvious. Located just off the Milky Way, the low power sweeping is great fun when the constellation is well placed during August nights.

Around April 22nd, after a long gap since the Quadrantids in January, approximately 15 meteors an hour will apparently emerge from the constellation, halfway toward the Hercules keystone.

Historically

A lyre is a seven-stringed musical instrument similar to a harp. In this instance, the instrument is deemed to be that which belonged to Orpheus, a fine musician. The lyre was an invention of Hermes, a son of Zeus, who fashioned it from a tortoise shell. Orpheus was a member of the Argonauts and, most famously, travelled to the Underworld to rescue his wife Eurydice from Hades, but failed at the very last moment.

Later myths from Arabia describe this compact group as a vulture or eagle.

Notable Stars

Double: Zeta Lyrae 4.3/5.9 and 44″. Yellow stars.

Double: Epsilon Lyrae A quadruple binary star where each star of a pair is itself a closely matched double. The pairs are stars of 4.7/6.2 and 5.1/5.5 magnitude separated by 208″. The close pairs are separated by just 2.6″ and 2.3″. The famous double–double. All are white stars.

Double: Beta Lyrae 3.5/7.2 and 46″. Blue stars. The primary star is a very close binary and mutual eclipses leads to the stars appearing variable.

Variable: RR Lyrae. Variables of this type pulsate in such a way as to make them a standard candle – though they are dimmer and vary faster than their better known Cepheid cousins. They are frequently found in globular clusters.

Deep Sky Objects

M57: The Ring Nebula. A classic planetary nebula, with an elliptical annular ring that can be spotted as a blurry dot in 10×50 binoculars. It is obvious in a 60mm refractor, with a 100mm showing its annular nature. The light from M57 is equivalent to that from an 8th magnitude star. The central white dwarf star is magnitude 14 and not easily seen. One of the most famous examples of a planetary nebula, exceeded, perhaps, only by the nearby M27. Essential viewing.

M56: An attractive and underrated globular cluster found halfway between beta Cygnii (Alberio) and gamma Lyrae. It may be resolved with a 250mm aperture. An old cluster, which may once have belonged to one of the Milky Ways' satellite galaxies.

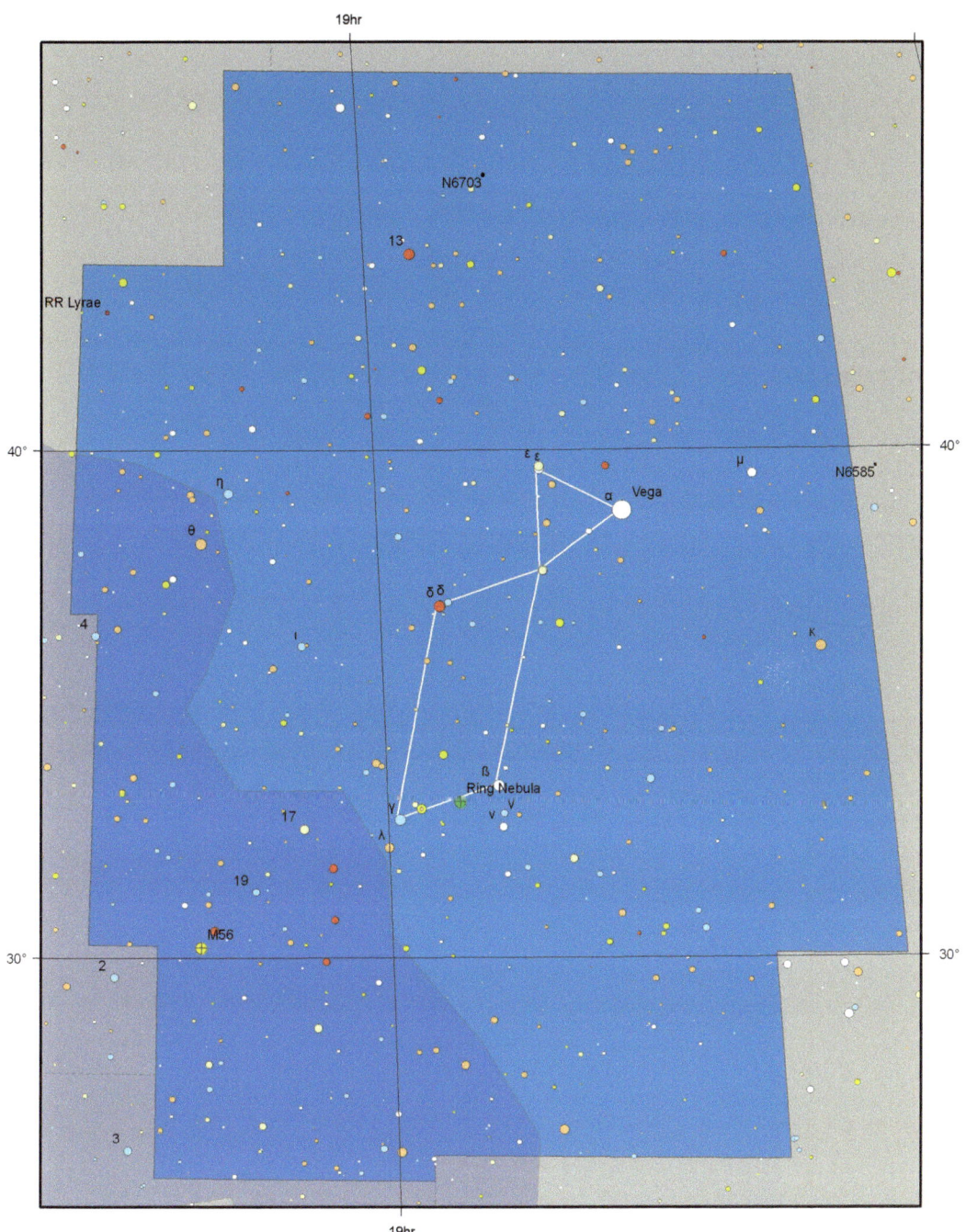

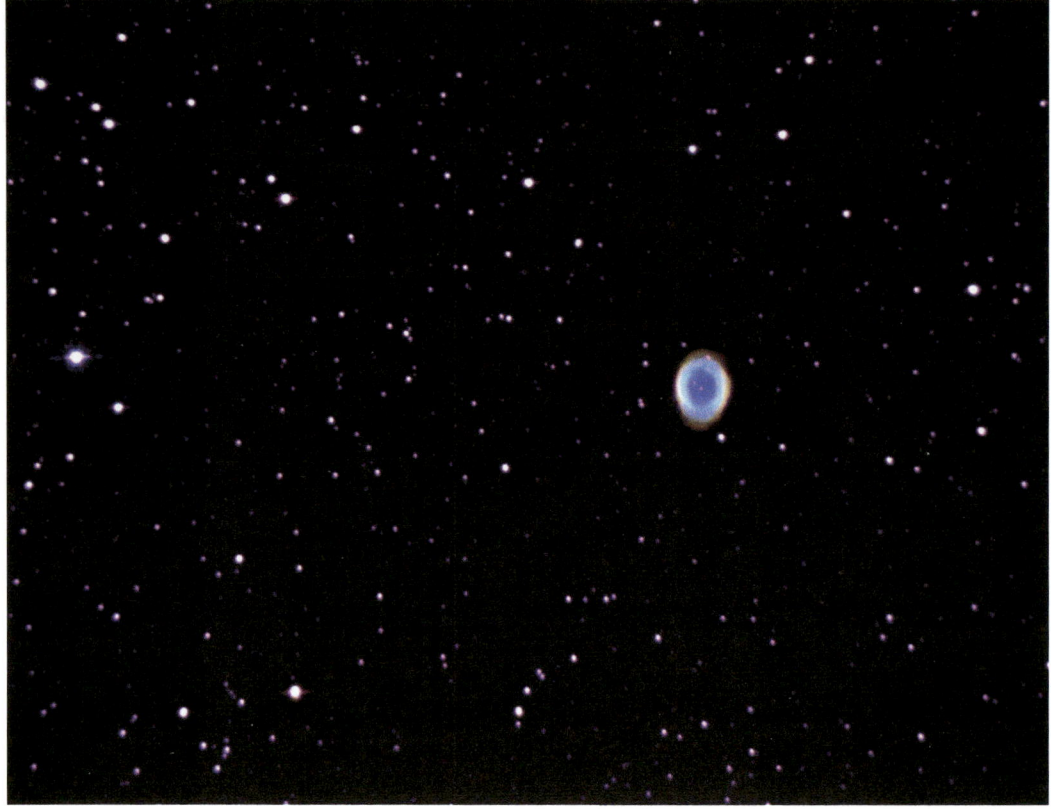

Fig. 17 The iconic Ring Nebula, M57, a disc-like planetary nebula in Lyra, is visible in an 80mm refractor (Image by Grant Privett)

MENSA

A southern hemisphere constellation, it has no bright stars and might be quite difficult to find, were it not adjacent to the Large Magellanic Cloud (LMC). Its brightest star is dimmer than every other constellations' brightest star.

It will not be easy to spot from urban areas but, from dark locations and the suburbs, it will be visible on Moonless nights at the start of the year.

Historically

To celebrate his stay in Cape Town, de Lacaille used a bunch of 5th magnitude stars to depict Table Mountain; with the LMC perhaps denoting cloud hanging over it. Mensa is Latin for "table". Originally the unimpressive grouping of stars was called Mons Mensae, but this was eventually shortened to simple Mensa.

Notable Stars

No double stars.

Sunlike star: Alpha Mensae is a Sun type star (class G5) of magnitude 5.1.

Deep Sky Objects

LMC: On the northern edge of Mensa the Large Magellanic Cloud sprawls into the constellation. Beside this there is little to observe in the deep sky, for smaller apertures. The cloud itself is, of course spectacular, even to the naked eye, and is impressive in binoculars.

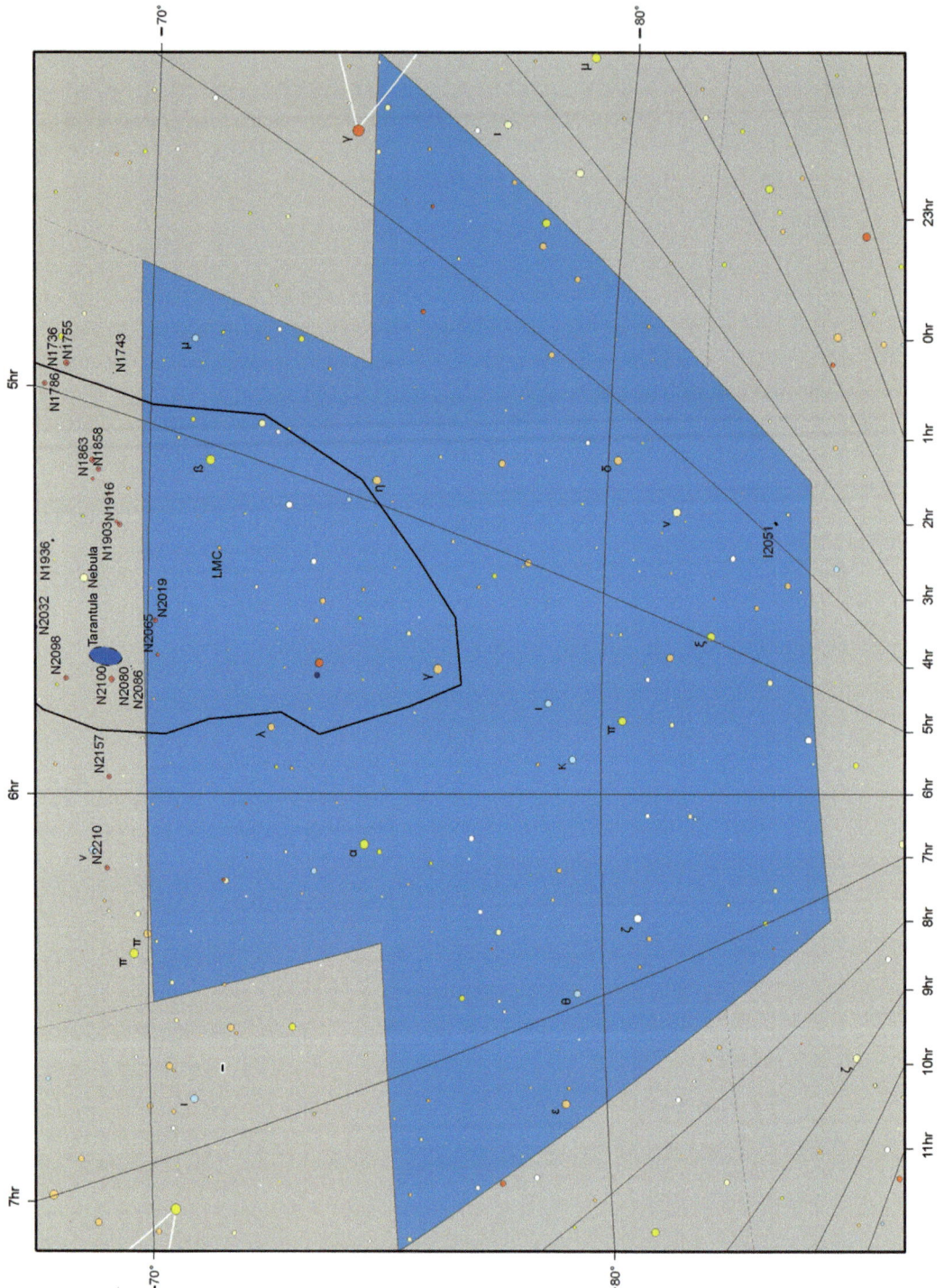

MICROSCOPIUM

A deeply unimpressive southern hemisphere constellation. It is certainly a competitor in the "most easily overlooked" constellation list, along with Octans and Mensa. A frequently-missed sight in the skies of September.

Historically

Microscopium commemorates the invention of the compound microscope by the Dutch spectacle maker Zacharius Janssen. At the time, it must have been as exciting as the invention of the camera or the internet. So, it's a bit of a shame that the constellation has no stars brighter than 5th magnitude.

Notable Stars

Double: Alpha Microscopii 4.9/9.8 and 20″. A line of sight double. The primary is strongly Sun like.

Variable: U Microscopii is a Mira type variable varying between magnitude 7 and 14 in about 330 days.

Deep Sky Objects

There are no notable deep sky objects in Microscopium.

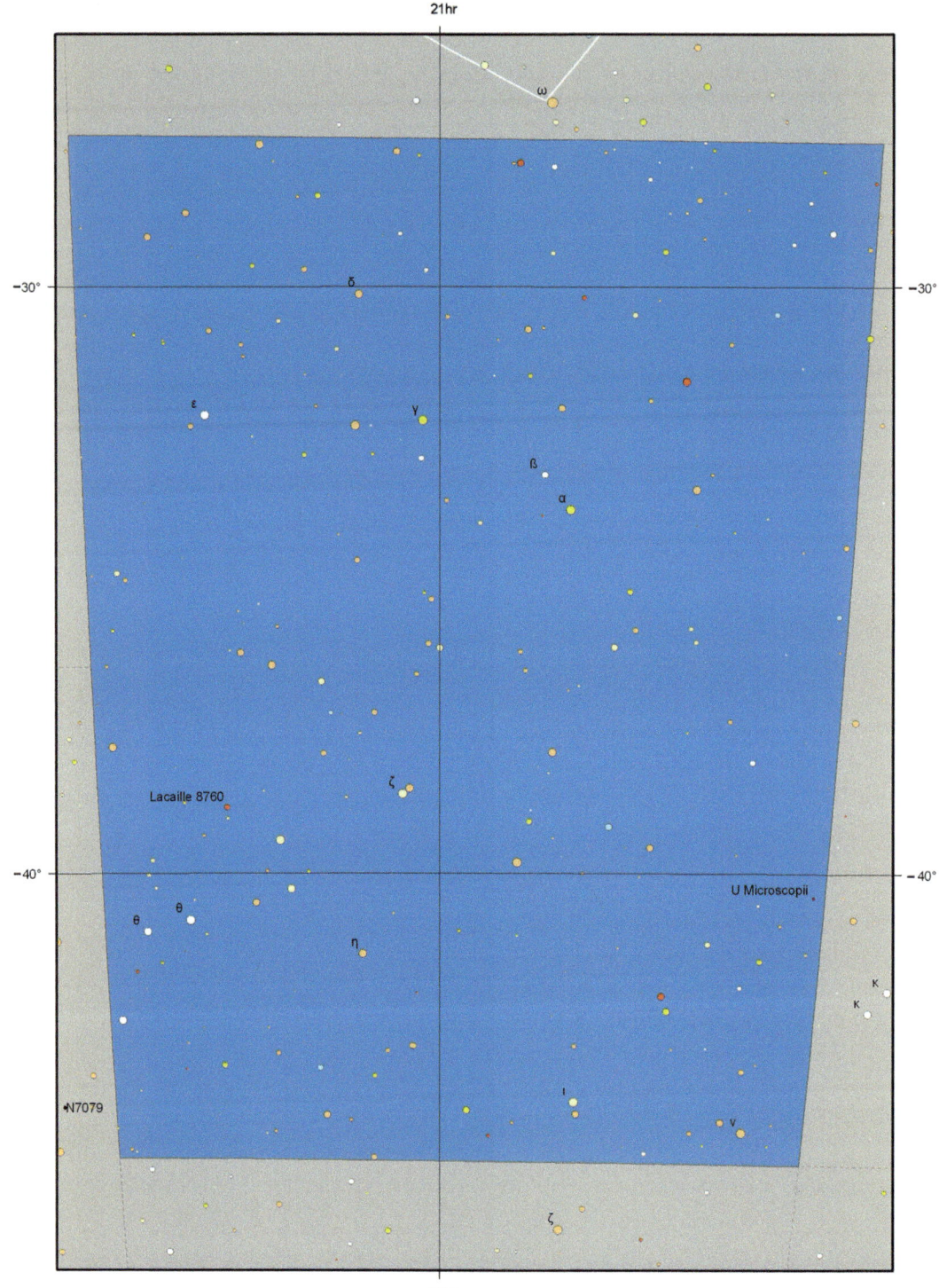

MONOCEROS

Monoceros is a rather sparse constellation of not very bright stars – its brightest, alpha Monocerotis, is just magnitude 3.9 – but it does sprawl across part of the Milky Way, which pretty much guarantees the constellation has some rather nice clusters and nebulae. The brilliant stars Sirius and Procyon bound it nicely.

Normally, there is no obvious meteor activity from Monoceros but, occasionally, short lived and intense outbursts have been detected. Keep an eye open for unexpected activity around 21st November.

Historically

Petrus Plancius invented this constellation, perhaps inspired by several Biblical references to the Unicorn. It's a bit dim for such a lovely creature, alas. Perhaps he was thinking of a rhinoceros.

A different, long forgotten constellation once filled in the slight gap between Canis Major and Monoceros. Jerome de Lalande and Johannes Bode created charts showing the printing press created by Johannes Gutenberg – Officina Typographica. A wonderful name for a constellation, and certainly a more worthy subject than some that have survived. But the stars really didn't look like a press at all and eventually the constellation fell by the wayside of astronomical history.

Notable Stars

Double: Beta Monocerotis 4.7/5.2 and 7.3″ with another at 10″. A wonderful triple star. Blue-white stars. A beautiful sight.

Double: Epsilon Monocerotis 4.5/6.5 and 13″. Yellow stars. A line of sight double.

Variable: U Monocerotis is a supergiant poised to shrug off its outer shells and become a white dwarf – an RV Tauri star in fact. It has two periods 92 days and 2,475 days.

Deep Sky

NGC2261: Hubble's Variable nebula. The brightest of the variable nebulae that change in appearance – the fine detail in the nebula comes and goes with time. The bright star associated, 9th magnitude variable star R Monocerotis is young and still forming. The gas swirling round it brightens and fades as the starlight is interrupted by clumps of dust orbiting close to R Mon. Ninth magnitude and just 2 arc minutes across.

M50: A bright young open cluster about two-thirds the width of the Moon. Visible in binoculars and small telescopes. A red giant is present at its core.

NGC2264: A bright open cluster, 4th magnitude, Incorporating a nebula called the Cone and 10–20 stars. Occasionally called the Christmas Tree cluster.

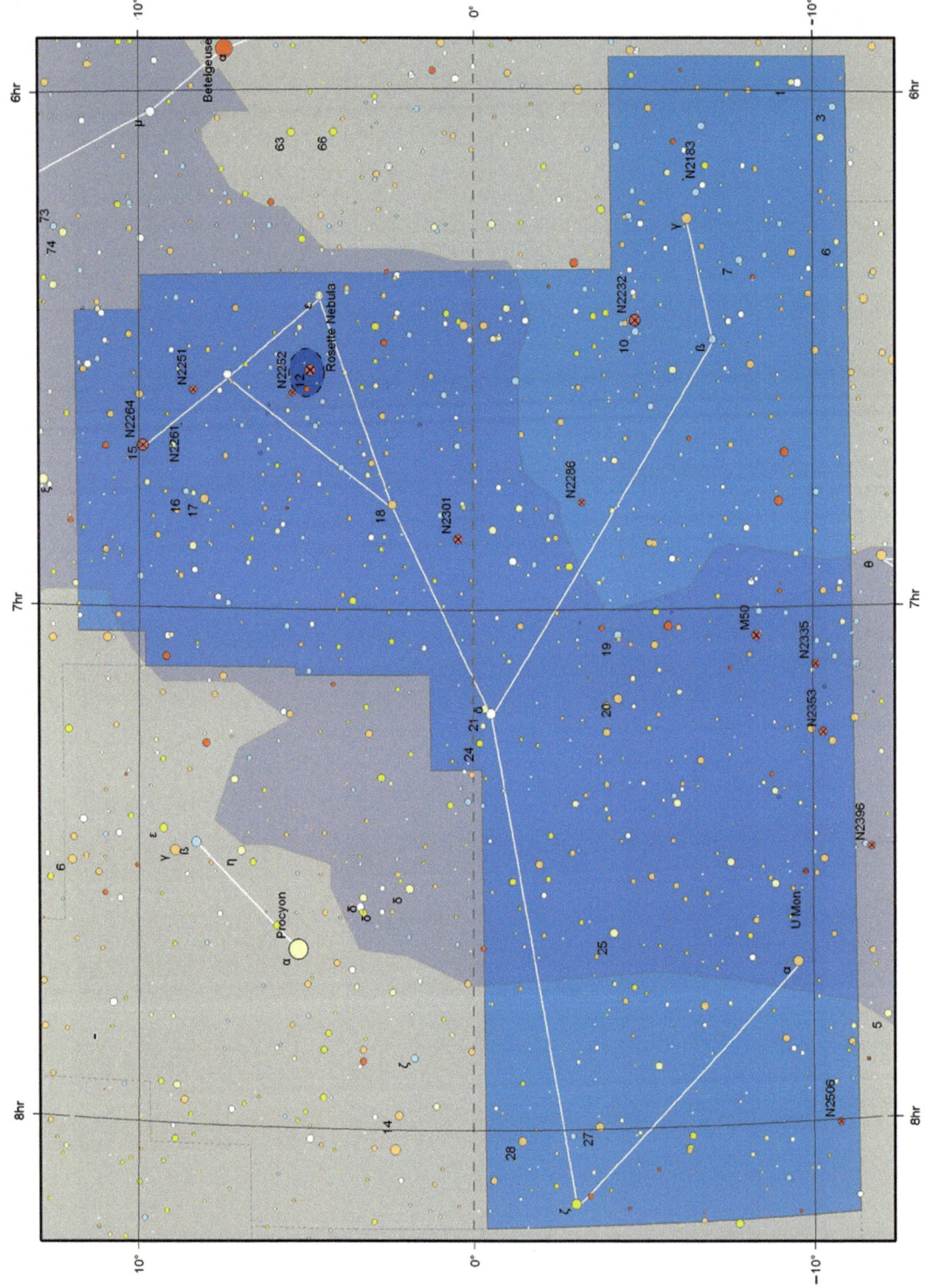

Fig. 18 NGC2261 is a nebula in Monoceros which changes its appearances over a period of just a few days (Image by Grant Privett)

MUSCA

A fairly small (just 138 square degrees) southern hemisphere constellation lying immediately south of the spectacular Crux. It lies just off the central plane of the Milky Way, but is still within its limits and is seen against a spectacular part of the sky. Its outline is quite conspicuous, if a little formless.

Look for Musca through April to June.

Historically

A grouping of stars first identified in the west by the travelers Houtmann and Keyser, who contributed so much to star charts. The stars were plotted by Plancius and eventually named the Fly, although until the constellations were formally described in 1930, it was sometimes known as Apis – the bee.

Notable Stars

Double: Beta Musca 4.0/4.0 and 1.6″. Very, very close but equally-bright blue stars. Worth trying to separate with webcam imaging

Double: Theta Musca 5.5/8.0 and 5.3″.

Deep Sky Objects

NGC4372: A globular cluster of 7th magnitude. It would probably be substantially brighter if it wasn't for a large amount of galactic dust absorbing much of its light.

NGC4833: A dim 9th magnitude globular cluster.

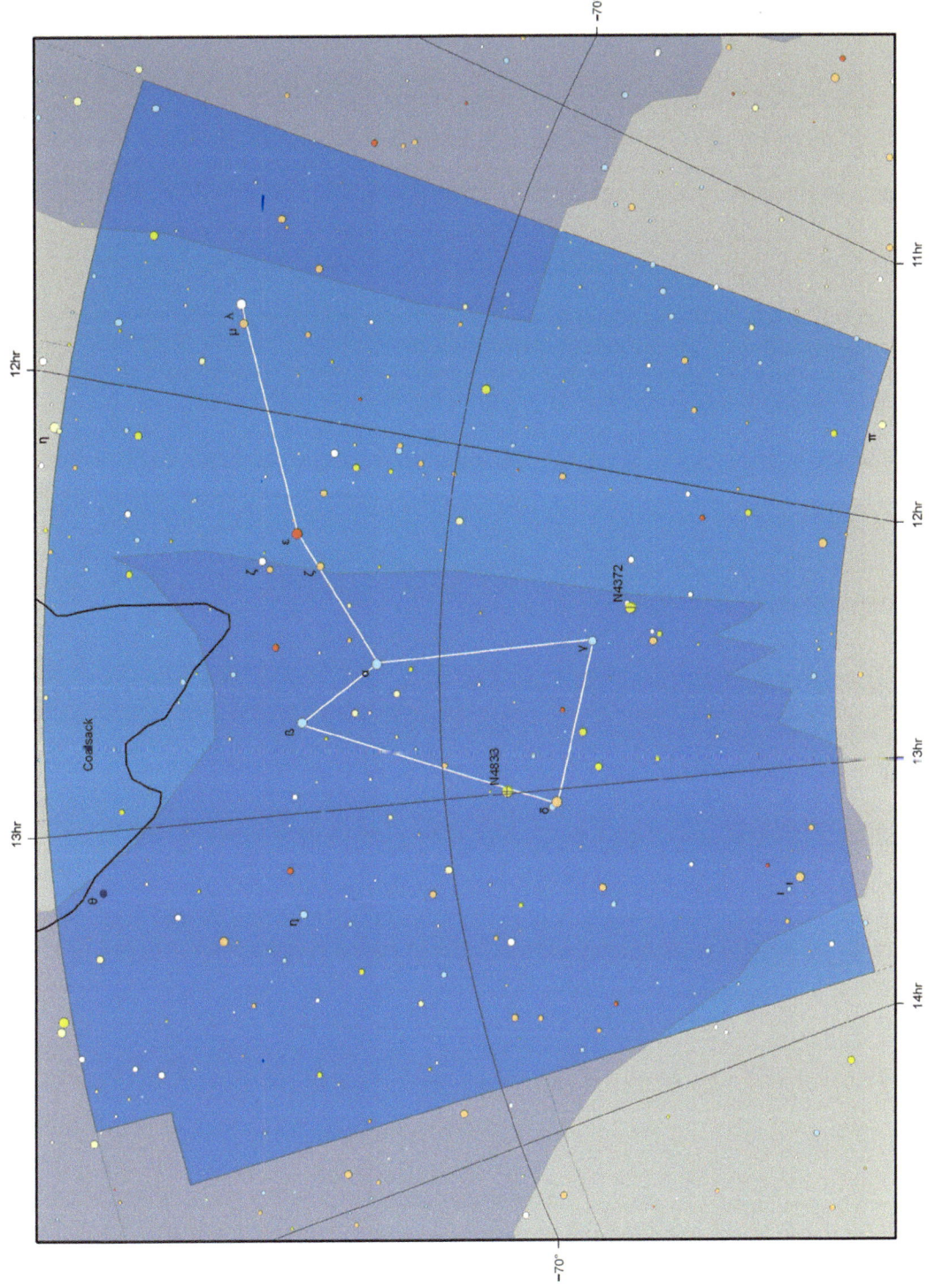

NORMA

A southern hemisphere constellation. One oddity of the constellation is that it lacks stars that are denoted alpha and beta Normae. This is because, when the constellation boundaries were fixed, the stars that filled these roles were transferred over to Scorpius.

It's fair to say that Norma isn't really anything to write home about, and may not even be visible in some suburban locations. Look for a "V" of 4th magnitude stars between alpha Centauri and the sting in the tail of Scorpius, during July.

Historically

The French astronomer de Lacaille had a good imagination, combined with a strong work ethic. While in Cape Town, at the southern tip of Africa, he catalogued over 10,000 stars with his small refractor and still found time to create 14 constellations – many of which were later embellished beautifully and imaginatively by Johann Bode in his Uranographia maps published in 1801. Norma actually represents a draughtsmans set-square and rule. It lies close to his constellation Circinus, which represents compasses.

Notable Stars

Double: Epsilon Normae 4.5/7.5 and 22".

Variable: R Normae is an unusual Mira type star exhibiting a double peak as it varies between magnitude 7 and 13. A fascinating object to follow

Deep Sky Objects

NGC6087: A bright but coarse open cluster of 5th magnitude. It only has about 40 stars but those are 7th magnitude or fainter – the exception being the Cepheid variable star S Normae at magnitude 6. Look out for S Normae's orangish appearance. NGC6087 spans 10 arc minutes, or thereabouts.

NGC6067: A compact and rich open cluster covering just 12 arc minutes. It is seen against a particularly bright portion of the Milky Way – the Norma Cloud. It has stars large and small and so boasts both youngish blue stars and older giants – including a super giant. In dark locations NGC6067 is visible to the naked eye. Well worth looking at with any instrument – it has 5× as many stars as NGC6087.

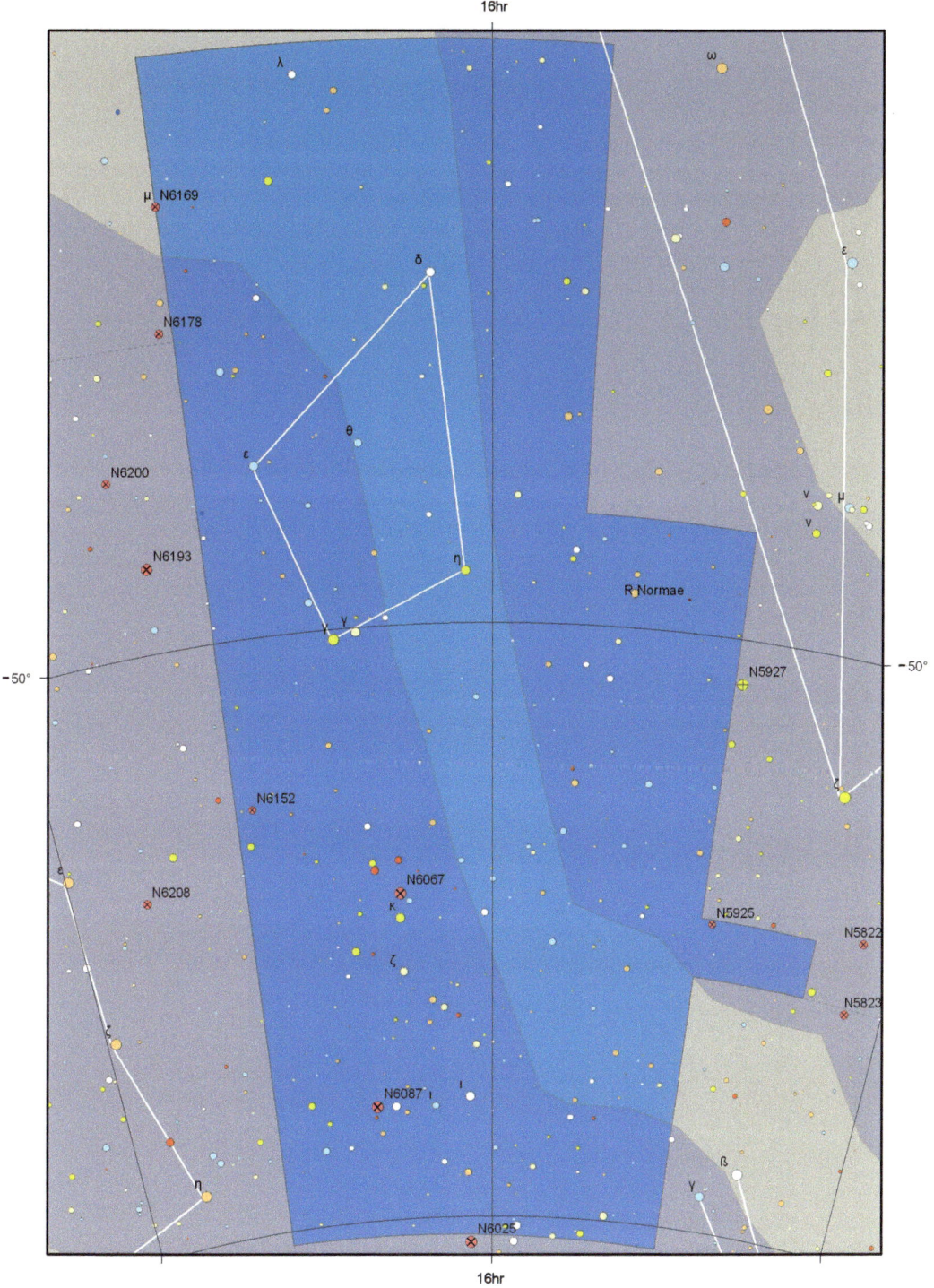

OCTANS

A faint constellation that encompasses the South Celestial Pole. Unlike the North Celestial Pole
– which is conveniently located close to Polaris – if you live in the southern hemisphere, there is
no bright star to make it easier to find your way, or to align a telescope.

The naked eye star nearest to the pole today is the highly unimpressive 5th magnitude star, sigma
Octanis. There's little of note about the star, apart from its position, and even that is over a degree
away from the pole. Precession is moving the pole away from sigma and, also, the North Celestial
Pole away from Polaris: it will be centuries before the situation improves much.

When looking for Octans, remember that its brightest star is of 4th magnitude. Avoid confusing
3rd magnitude beta Hydri for beta Octanis – as I did the first time I saw the constellation.

Historically

Anyone who has read our notes on the more obscure and unsung southern constellations, will have real-
ized that when a constellation was created by de Lacaille it was unlikely to have anything to do with
mythology and far more likely to be a bit of contemporary hardware. Here it's true again. Octans represents a reflecting octant. They are not much used these days – a strange gap in Apple's inventory – as
they are a navigational instrument predating the sextant. The octant was invented by John Hadley in
1730.

Notable Stars

Double: Lambda Octanis 5.5/7.6 and 3.0″.

Deep Sky Objects

There are no notable deep sky objects in Octans.

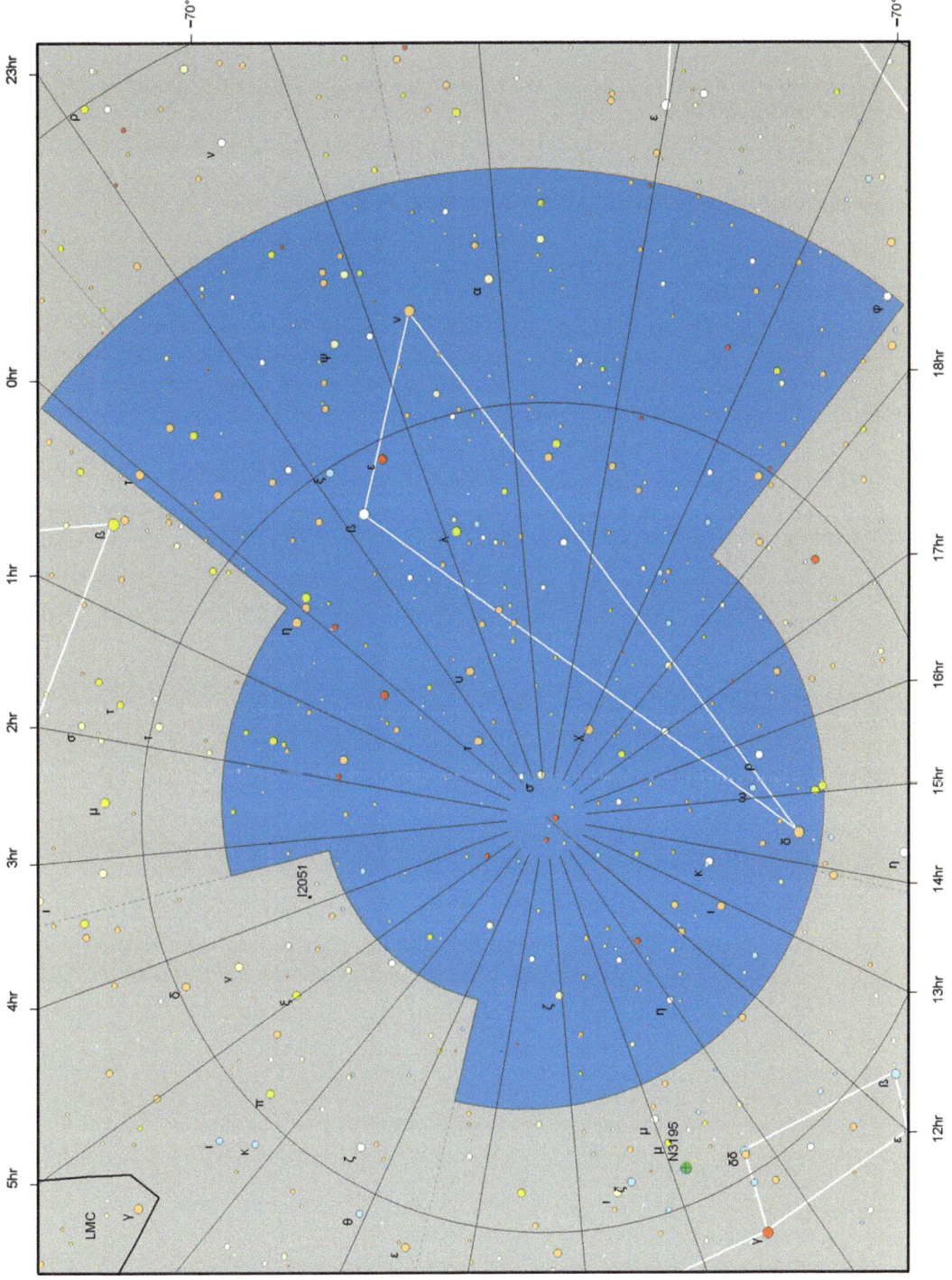

OPHIUCHUS

A constellation straddling the celestial equator and thus visible from most of the occupied parts of the Earth. The constellation sweeps across 43° of declination and, at its southern extreme, includes a small part of the ecliptic. Ophiuchus would be the thirteenth sign of the zodiac, if astrologers were in touch with reality.

It's *the* constellation to visit if you like observing globular clusters.

Historically

This is another constellation that is vaguely man shaped. In this case, it represents a figure grasping the two halves of a snake, or grasping the head and tail, with the rest of the snake wrapped around him. The stars making up the snake appear under the constellations Serpens Cauda and Serpens Caput, (Snakes' Tail, and Snakes' Head, respectively)

Notable Stars

Double: Omicron Ophiuchi 5.4/6.9 and 10.3″.

Double: Xi Ophiuchi 4.5/9.0 and 3.7″.

Double: Rho Ophiuchi 5.3/6.0 and 3.1″. Blue B2 type stars.

Deep Sky Objects

M10: A superb binocular globular cluster. Sixth magnitude and large, lying within a rich field of view. Not quite spherical, with a reasonable amount of condensation towards its core. Takes a telescope in the 300mm league to start being resolved.

M12: A fine globular cluster that can be resolved in a 200mm aperture. It is quite a loose cluster and appears as a 7th magnitude patch of gossamer, perhaps 5 arc minutes across in smaller instruments. Not as bright as M10.

NGC6633: A 4th magnitude cluster just 20 arc minutes across, lying near the border. It has quite a few bright members.

NGC6572: A tiny planetary nebula of around 15×7 arc seconds diameter (about the apparent width of Mars at opposition). The central star is 11th magnitude and the nebula 9th, so it will be seen even in small instruments. Some observers report it as strongly colored – greenish or blue. The HST image of it is fascinating.

IC4665: A loose and sparse open cluster of 4th magnitude. It is quite large, at nearly a degree across, and well seen in binoculars and smaller telescopes. A 100mm should spot 30, or more, members.

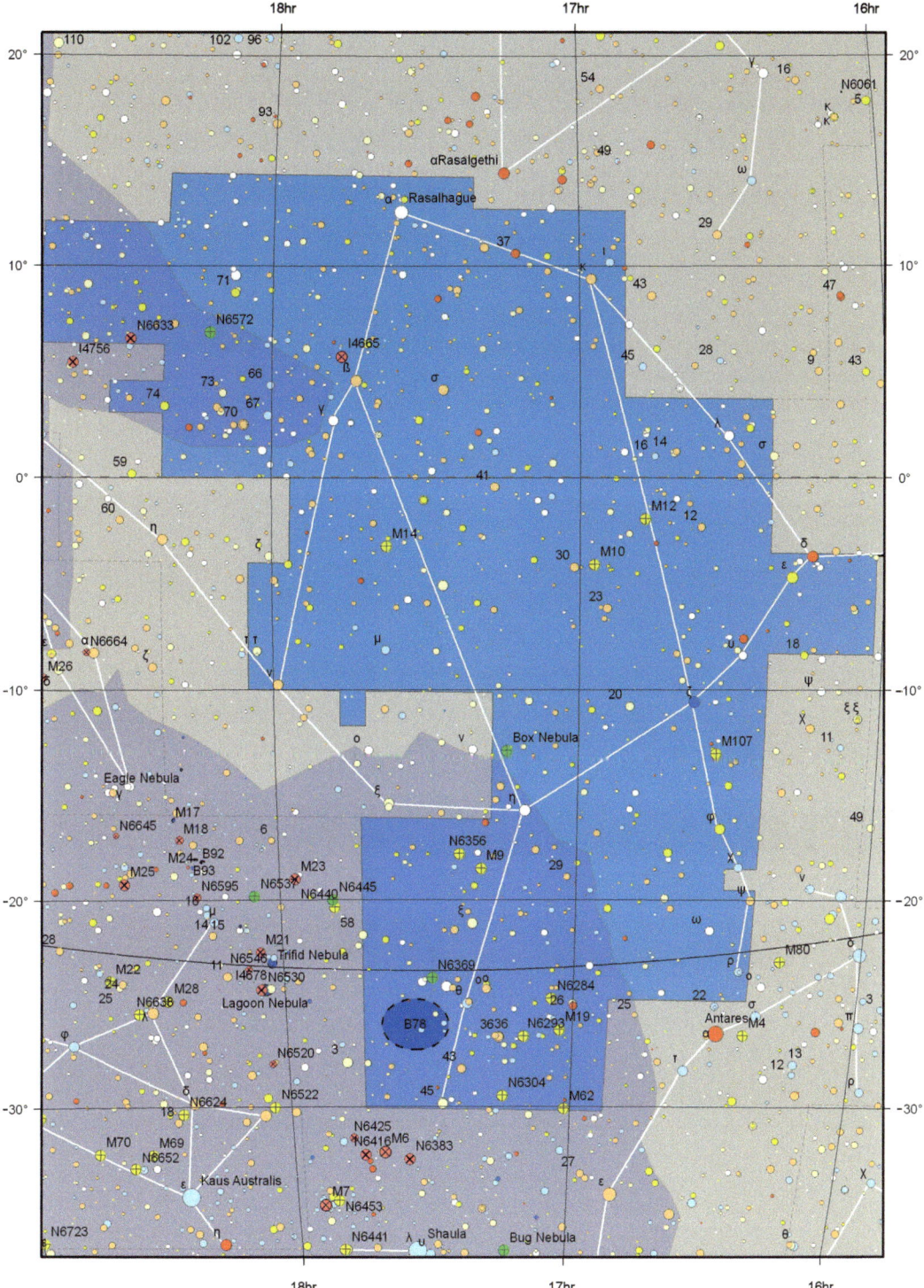

Fig. 19 M10 is a bright globular cluster in Ophiuchus. A number of similar clusters are visible in Serpens (Image by Bill Snyder)

ORION

One of the sky's great constellations. A lovely sight in the evenings of December and January. It lies on the edge of a region of the Milky Way looking away from the galactic core but, even so, boasts some of the most glorious nebulae within the sky. On a clear night it's almost impossible to miss – one of the first constellations a new astronomer learns.

A meteor shower, the Orionids, appears in the north west of the constellation during the week or so either side of October 21st. As many as 25 meteors an hour may be seen under ideal conditions.

Historically

As a constellation that looks so very much like a man, it's no surprise that these stars are viewed as being those of a hero. In this instance, a great warrior, with a shield on his arm and a belt and dagger at his waist. In Greek mythology he is deemed to be the huge warrior who was killed by a scorpion – represented on the sky by the constellation Scorpius – and also the one who forced his attentions on a princess named Merope. Needless, to say the Babylonians, Egyptians and Romans all had their own distinct heroes in mind.

Notable Stars

Double: Delta Orionis 2.2, 6.3 and 52.7″. White and bluish stars.

Double: Lamda Orionis 3.8/5.6 and 4.4″. White stars.

Double: Theta Orionis (The Trapezium) 6.7/7.9/5.1/6.7 and 8.8/13/21.5″. A quadruple star at the heart of the orion Nebula. A must see object.

Double: Beta Orionis 0.1/6.8 and 9.5″. Blue and white stars.

Double: Iota Orionis 3.0/7.0 11.4″. White and blue stars. Lies within NGC1980

Double: Sigma Orionis 4.0/7.5/10/6.5 and 12.9/11.2/42.0″. Lovely quadruple star. The dim star is probably a member.

Double: Zeta Orionis 2.0/5.5 and 3″. Bluish and yellow.

Variable: Alpha Orionis, Betelgeuse, a semi-regular variable first noted by John Herschel. It varies between magnitude 0 and 1.3.

Deep Sky Objects

M42/M43: Simply the most impressive nebula visible in the northern hemisphere. Wide, bright and with a quadruple star at the core. A superb sight. Even a 130mm will show sufficient detail to make drawing it challenging. A must-see object.

M78: Two closely spaced 10th magnitude stars encased in a mass of gas over 7 arc minutes across. Images show a wealth of detail and several smaller patches of gas and dust nearby. Lovely, detailed and often overlooked.

NGC1977: The Running Man nebula. A small cluster of brightish stars encapsulated in a dim reflection nebula in the same low power field as M42.

NGC2024: A dim, large and impressive nebula crossed by dust lanes.

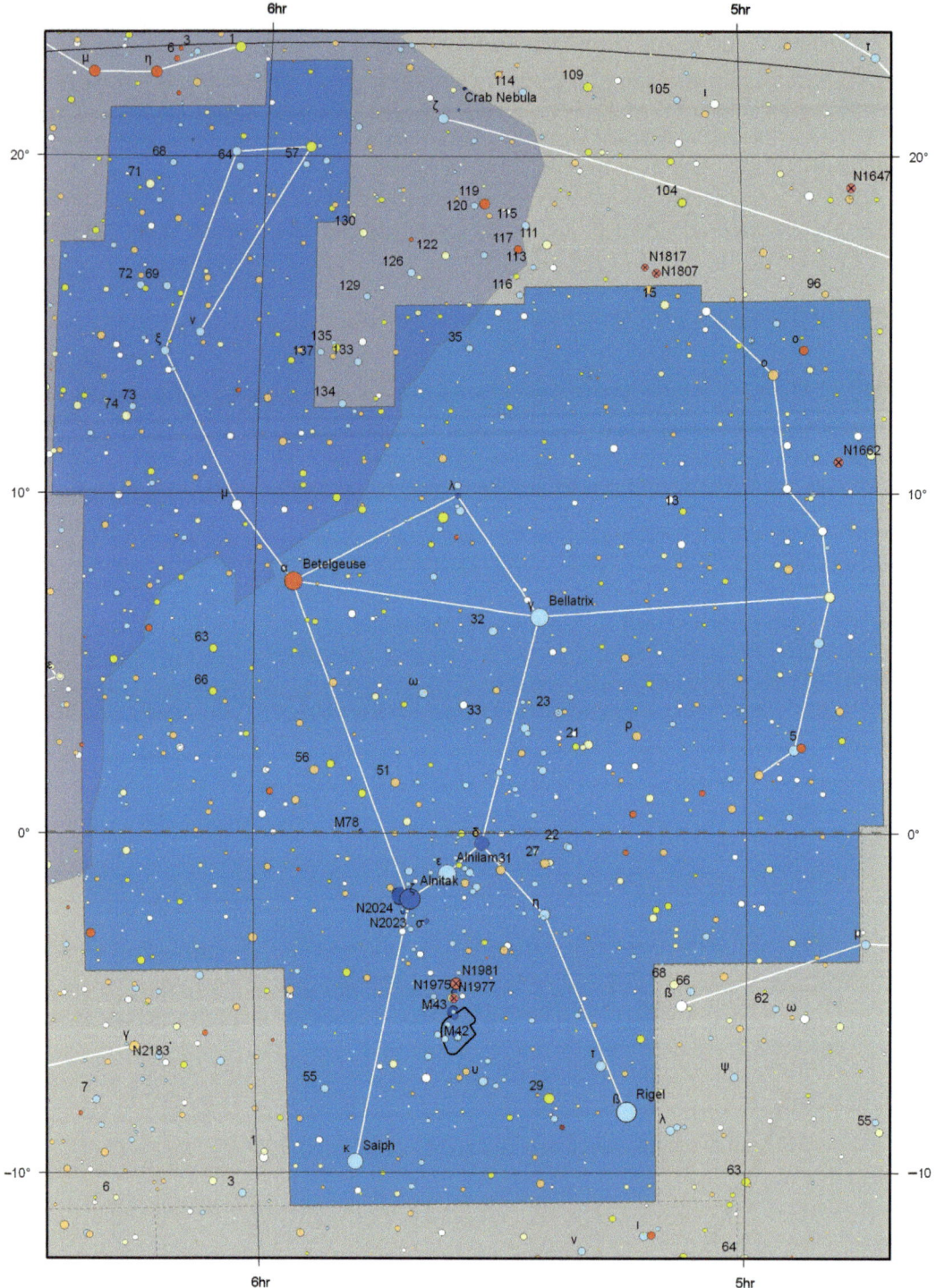

Fig. 20 A close up of the Orion Nebula, M42, and its companion nebula M43, which can be found in the Sword of Orion (Image by Grant Privett)

PAVO

A southern hemisphere constellation formed mainly of 3rd magnitude stars. It's close to the Milky Way and in a region where browsing the star field can be rewarding. Pavo is circumpolar from much of the southern hemisphere and so always on show. Its brightest star is the bluish alpha Pavonis – oddly named the Peacock – which shines at magnitude 1.9.

It will be well seen in August.

Historically

A constellation introduced by the chart making Petrus Plancius – one of the founders of the Dutch East India Company. It's notable that Argos, who built the ship Argo was, after his death, transformed into a peacock. This may be what Plancius was thinking of.

Unfortunately, the constellation doesn't do its name credit. A peacock is large bright and showy: this constellation certainly is not. It has, however, survived better than some of the constellations suggested by Plancius – such as Euphrates Fluvius et Tigris Fluvius, or the rivers Euphrates and Tigris – which never really caught on, despite the catchy name.

Notable Stars

Double: Xi Pavonis 4.5/8.0 and 3.5″. Yellow stars.

Sunlike star: Delta Pavonis has a spectrum very similar that of the Sun. Its color and metallicity – a measure of the amount of material present, that is not Hydrogen, Helium or Lithium – show a striking similarity to the levels found in our star.

Deep Sky Objects

NGC6752: A splendid 5th magnitude globular cluster covering a one-third of a degree of sky. Its brightest members are of magnitude 13th, while a 7th magnitude star lies close nearby, mingling with the clusters outliers'. At ten billion years old, you would expect the stars to be old and yellow but, like many globular clusters, it contains "blue straggler" stars which resulted from mergers of smaller stars in the close confines of the cluster cores. Contains 100,000 stars.

NGC6744: A rather nice inclined spiral galaxy; a bit like a cross between M81 and M101. Ninth magnitude and 20×13 arc minutes in size, it has many HII – star forming – regions but only the core will be seen in smaller instruments. Averted vision will help locate the dimmer outer regions.

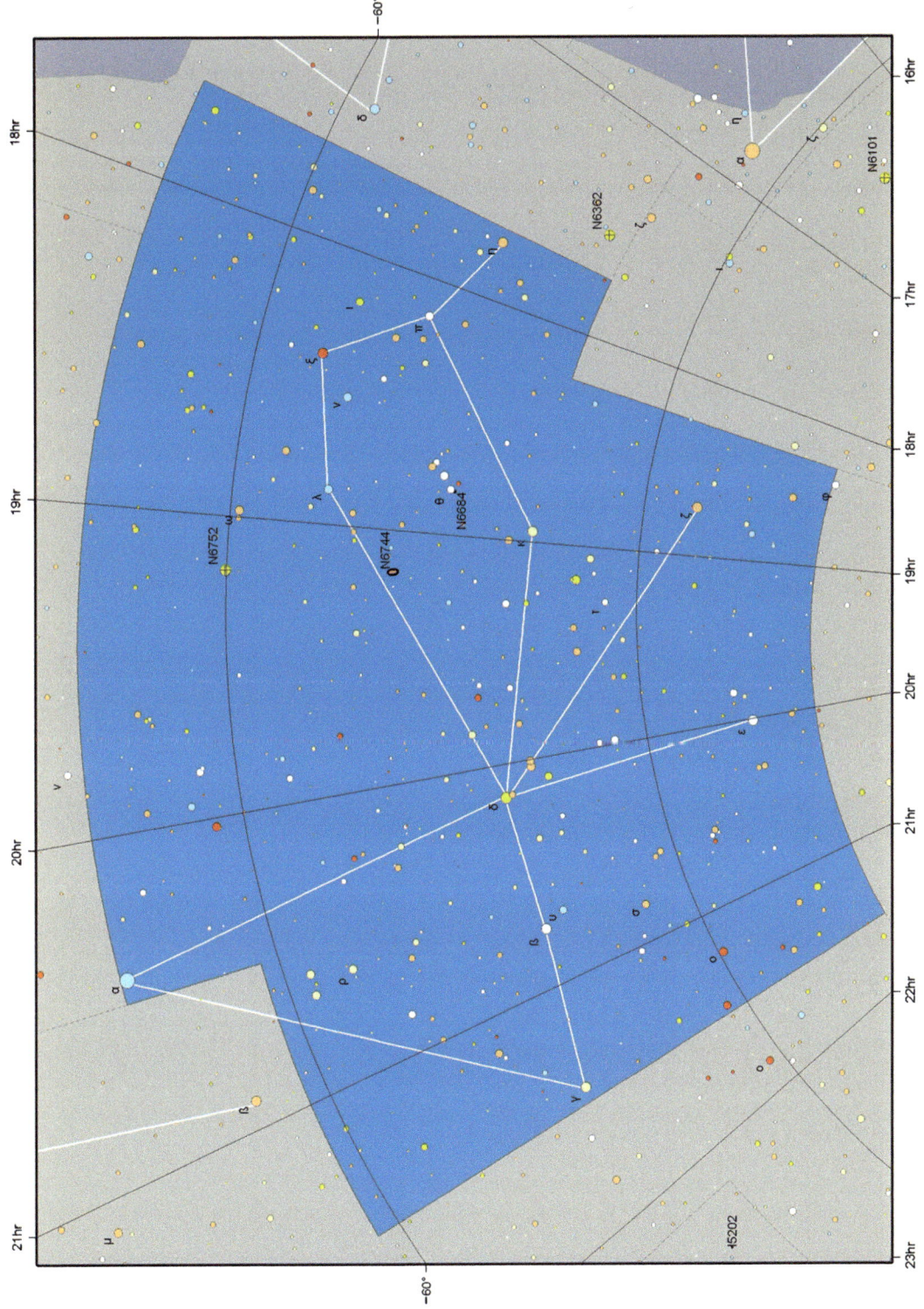

PEGASUS

A large northern hemisphere constellation dominated by the Square of Pegasus – the shape made up from the 2nd magnitude stars alpha, beta and gamma Pegasi, plus alpha Andromedae, which was originally a star within Pegasus.

The region within the square is fairly sparsely populated, and is frequently used to test the limiting magnitude. See how many stars you can see within the square. If it's more than three stars, you are seeing down to 5th magnitude which is quite respectable for the light polluted UK.

Pegasus is the 11th largest constellation.

Historically

The birth of the winged horse, Pegasus, is one of the stranger Greek legends. Pegasus was born as Medusa was slain by Perseus. It's all a bit strange: Medusa was not always ugly and was only turned into the hideous snake-haired monster after she was seduced by Poseidon. The seduction took place within the Temple of Athene, who was naturally miffed. Typically of mythology, no blame seems to have attached to Poseidon. I suppose that's the virtue of having godly powers of one sort or another.

Later, Pegasus assisted the hero Bellerophon on his quest.

Notable Stars

Double: Epsilon Pegasi 2.5/8.4/11 142/82″. White or yellow stars. Triple star.

Double: Eta Pegasi 3.0/9.0 and 91″.

Variable: Beta Pegasi is a semiregularly varying reddish giant star with a period of 43 days. Range magnitude 2.3–2.7.

Deep Sky Objects

NGC7331: A 10th magnitude inclined spiral. Visible in a 150mm as a dim elongated blob, growing to 10 × 3 arc minutes in larger scopes. In images a number of satellite galaxies lie close nearby and to the east.

M15: A large, bright and rather lovely globular cluster visible in binoculars. It is 6th magnitude and around 8 arc minutes wide. Look out for it to be resolved in a 200mm and for there to be a sharply condensed core. One of the few globular clusters to contain a planetary nebula. Don't bother looking for it without a *very* large telescope, good finder charts and an OIII filter.

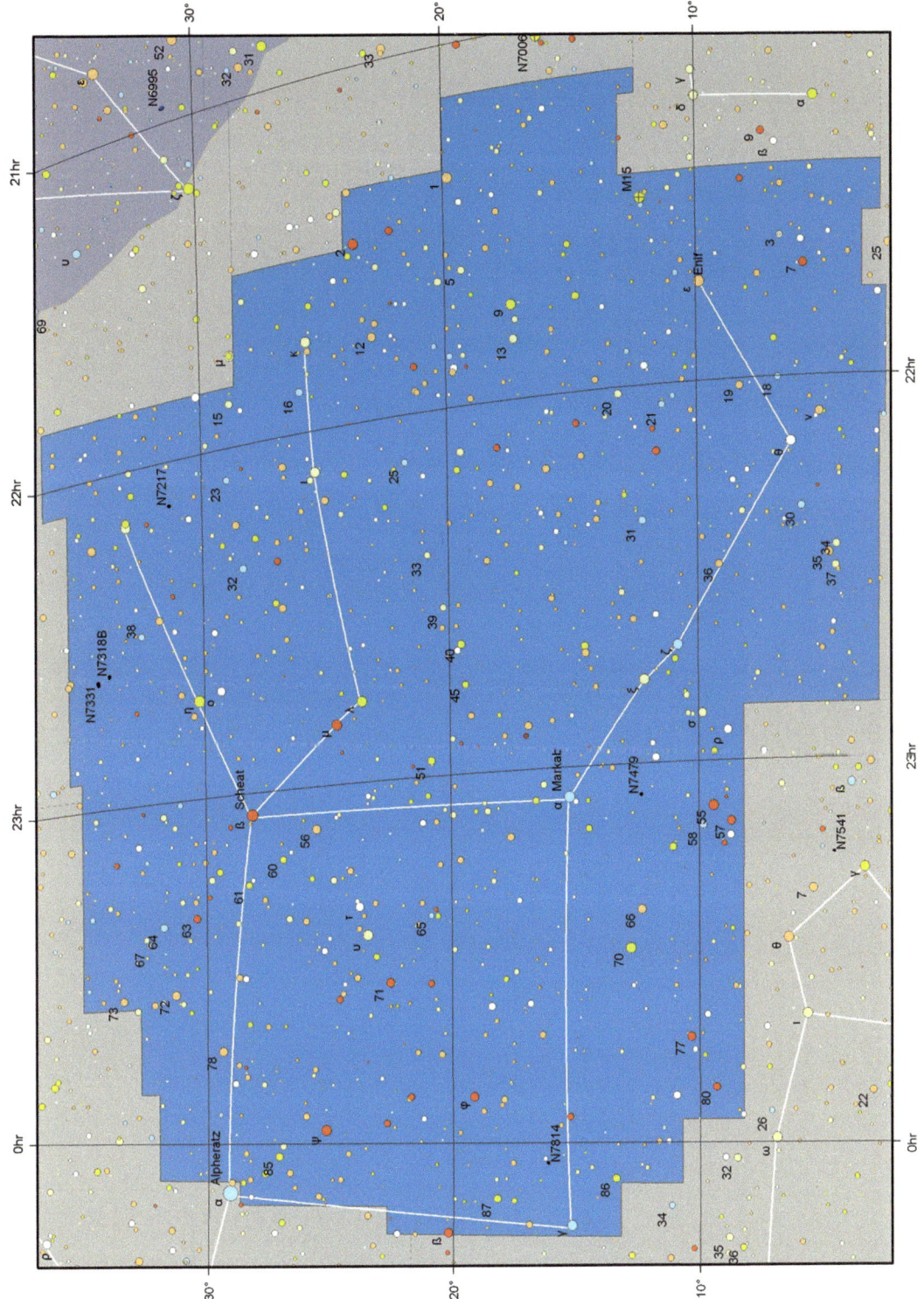

Fig. 21 NGC7331 in Pegasus, is an inclined spiral galaxy with several satellite galaxies. It is visible in a 100mm reflector (Image by Grant Privett)

PERSEUS

A northern hemisphere constellation that is only partially circumpolar from a north American/ European latitude. It is home to the impressive Perseid meteor shower which emnates from its north. The Perseids are active for a week or two, either side of the maximum on the glorious 12th of August. On the night of the peak, the hourly rate can reach 100 meteors (if the observing site is ideally located and is very dark). Another, much weaker, shower occurs on September 9th – but is a pale reflection of the August outburst.

Perseus is an obvious inverted "y" shape just below the "w" of Cassiopeia.

Historically

Perseus is a hero. Well, he would be, he is the son of Zeus! He was sent by his stepfather to kill the Gorgon, Medusa. A monster so hideous, that just looking at her would turn you to stone. With a little help from Mercury, *et al.*, he succeeds, and also gains the flying horse Pegasus along the way.

The flying steed comes in handy when he finds the maiden Andromeda tied to a rock, by her caring parents Cepheus and Cassiopeia. She is awaiting death in the jaws of a sea monster to placate Poseidon, the rather bad tempered King of the sea. Perseus saves her, and they escape. It's pretty exciting stuff and has been made into a film several times.

Notable Stars

Double: Eta Persei 3.9/8.5 and 29″. Blue and yellow stars.

Double: Theta Persei 4.1/9.9 and 20″. Yellow and white stars.

Double: Epsilon Persei 3.0/8.0 and 8.8″. Blue stars.

Variable: Beta Persei – Algol – the demon. A close double star where one component regularly eclipses the other, producing two dips in its normal brightness of magnitude 2.1.

Deep Sky Objects

NGC884/869: The Double Cluster. Two bright and large clusters lying 30 arc minutes apart. To fully appreciate them you will need a wide-field instrument, operating at a low power, to capture as many of their 500 total stars as possible. Obvious to the naked eye as a diffuse haze-like patch and well seen in binoculars. Ideal for a 114mm aperture reflector or 80mm refractor.

M34: A bright and distinctive cluster that is easily found with binoculars and hovering on the verge of naked-eye detection in dark skies. Its 50 or so brighter members will fill a 30′ field. In small scopes the 'X' shape made by the brighter members is obvious.

OB Association: The region around alpha Persei contains a grouping of young blue and white stars that make binocular and low power telescope views attractive – as if the Milky Way passing through wasn't a good start.

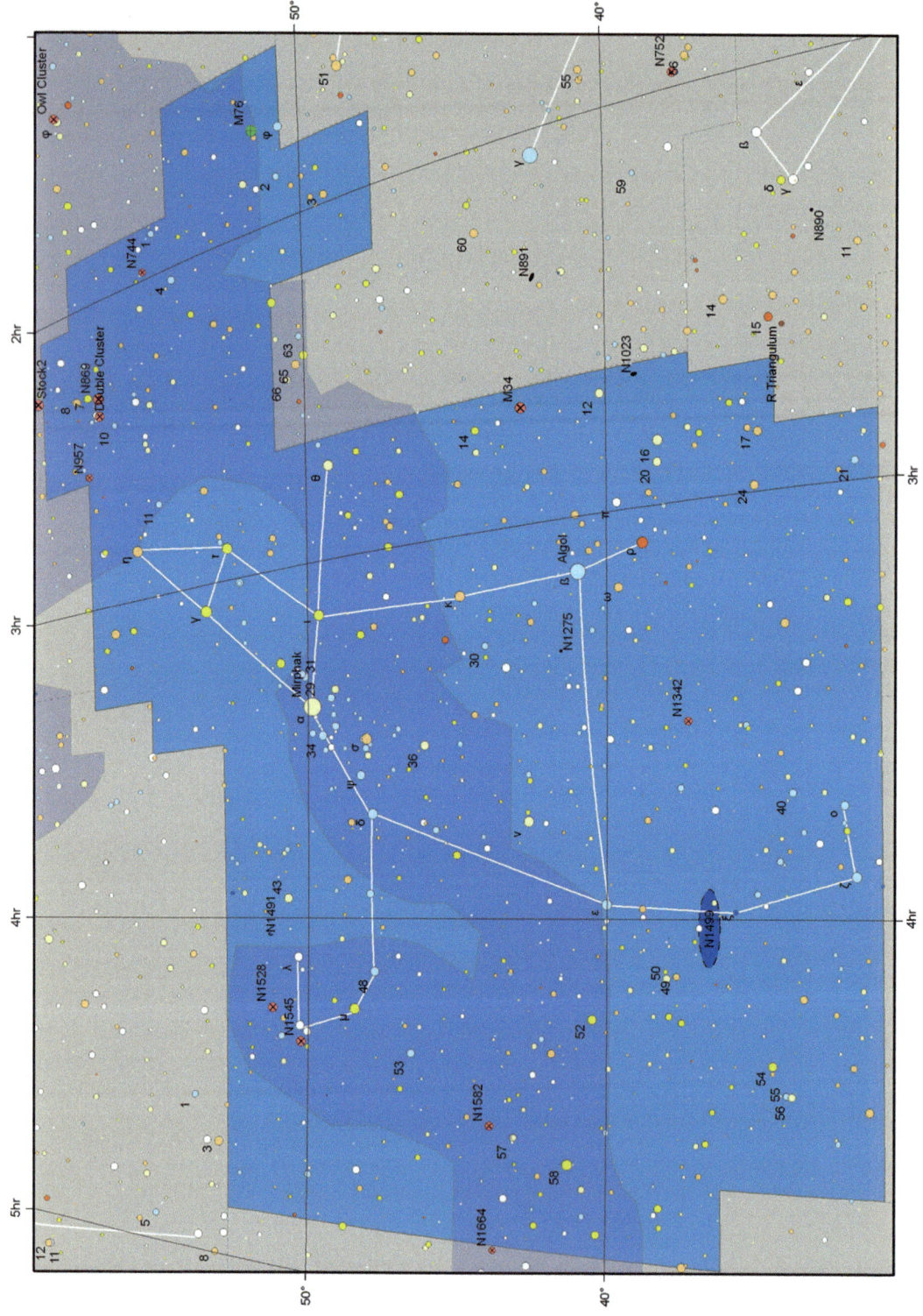

Fig. 22 The Double Cluster lies between the southern edge of Cassiopeia and Perseus and is discernible to the naked eye (Image by Bill Snyder)

PHOENIX

A southern hemisphere constellation lying in a less than impressive region of sky, which is relieved only by the nearby 0.4 magnitude star Achernar. The constellations' brightest star is 2nd magnitude and a couple more are 3rd magnitude.

In 1956, strong meteor activity was recorded from the Phoenix on the 6th December. It may happen again sometime. Keep a look out.

Look out for the Phoenix toward the years' end.

Historically

A creation of Petrus Plancius, this rather dim conglomeration of stars is deemed to represent the mythological phoenix bird rather than the colleague of Achilles. It celebrates rebirth and, perhaps, symbolizes the passing of the seasons.

Notable Stars

Double: Theta Phoenicis 6.5/7.0 and 4.0″. Both white.

Sunlike star: Nu Phoenicis is a Sunlike star complete with hints of its own asteroid belt – observed by infrared imaging.

Variable: Zeta Phoenicis is an Algol type eclipsing variable that varies over the range magnitude 3.9–4.4 in 1.7 days. It is also a double star with an 8th magnitude component lying just 6 arc second away.

Deep Sky Objects

There are no notable deep sky objects in Phoenix.

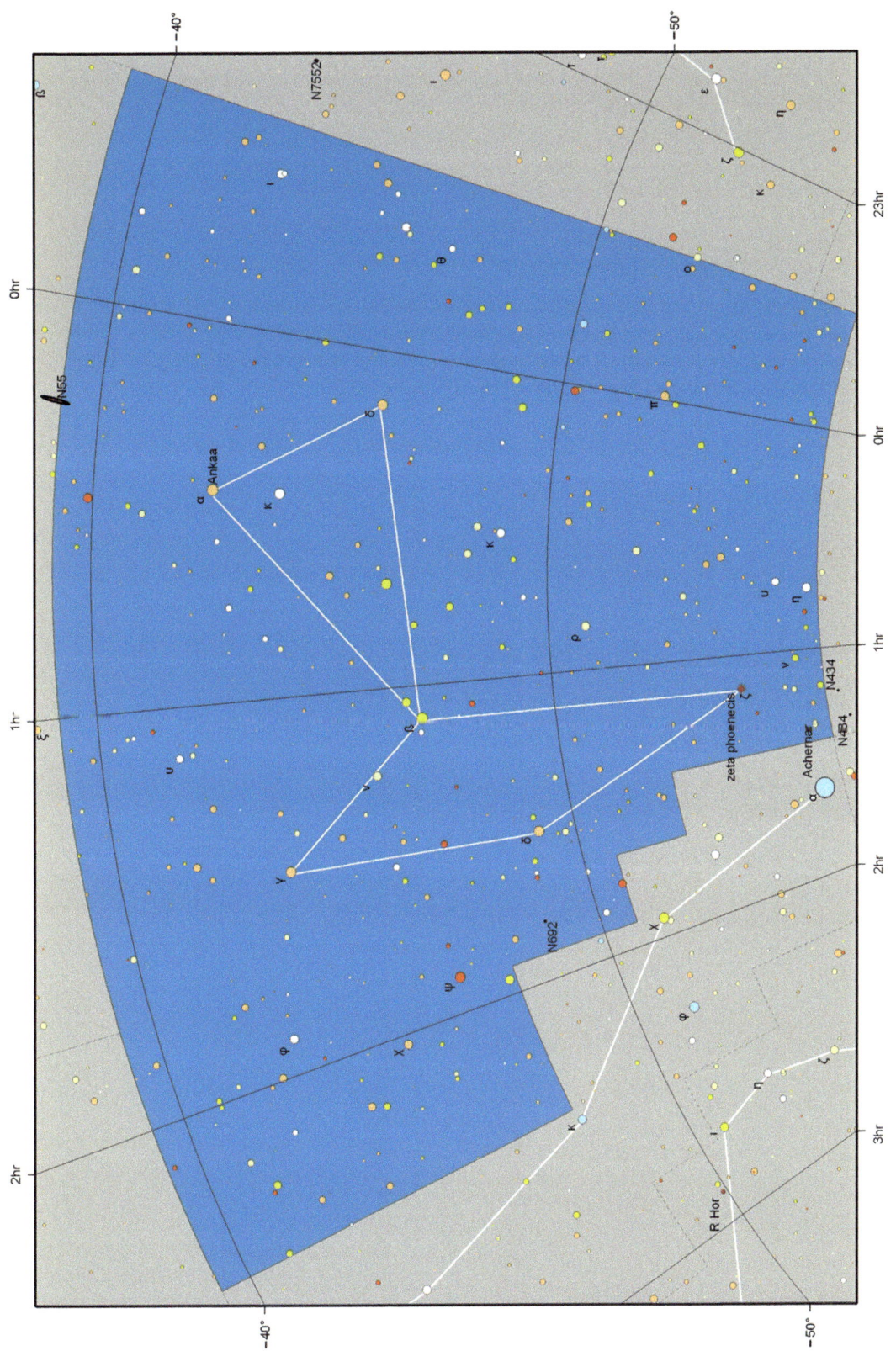

PICTOR

An unassuming southern hemisphere constellation of 3rd and 4th magnitude stars, lying close to the bright star Canopus. It is unlikely to get much attention, due to the rich fields of Carina and the Large Magellanic Cloud (one of our galaxy's satellite galaxies) in Dorado.

It will be most easily seen at the start of the year.

Historically

The painters easel and palette is a constellation named by de Lacaille in the 1750s. Originally it was le Chevalet et la Palette, (the Foal and Palette…it sounds like a theme pub with pretentions) but by 1763 he had converted that to the Latin Equuleus Pictorius. This was later abbreviated to Pictor to avoid confusion with the existing Equuleus constellation.

Notable Stars

Double: Iota Pictoris 5.5/6.5 and 12″. Yellow stars.

Variable: R Pictoris is a giant star and a semiregular variable star that spans the range magnitude 6 to magnitude 10 in about 170 days.

Deep Sky Objects

There are no notable deep sky objects in Pictor.

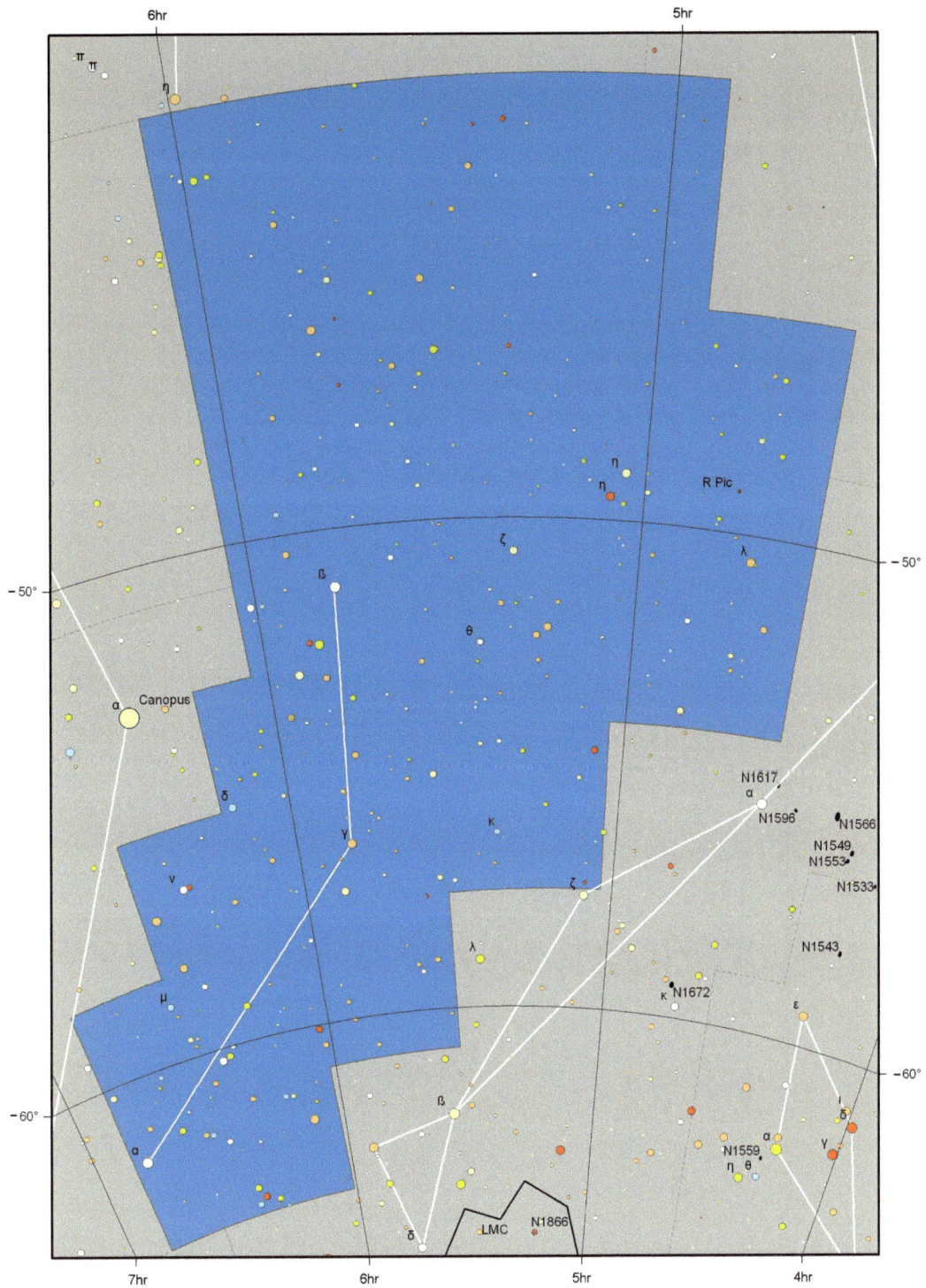

PISCES

A constellation lying across the celestial equator. Most of it lies north of the equator, but a small portion is southerly. The constellation also contains the path of the ecliptic and, so, is no stranger to the presence of bright planets. During March the Sun moves east from Aquarius and, passing along the ecliptic, crosses the celestial equator from southern to northern hemisphere. The date on which that occurs is called the Spring Equinox and marks the start of the northern hemisphere spring.

The outline of Pisces itself is not hugely interesting, but it is quite large – covering 247 square degrees.

Historically

A constellation depicting two fish tied together by a cord. In response to some very strange godly goings-on, the Goddess Gaia sent a hideous monster, the typhoon, to terrorize Pan and a number of other Greek gods. The fishes carried Eros and Aphrodite to safety, or were, possibly, Eros and Aphrodite transformed.

Notable Stars

Double: Zeta Piscium 4.2/5.4 and 24″. White and pale blue.

Double: Alpha Piscium 4.2/5.1 and 1.7″. Bluish stars.

Double: Psi1 Piscium 5.6/5.8 and 30″. Blue stars.

Deep Sky Objects

M74: The only notable deep sky object in Pisces is a large low surface brightness face-on spiral galaxy. Many a Messier marathon has stumbled when it came to finding this object. Overall, it is of 10th magnitude, but the light is spread out over 10 arc minutes – so it seems considerably fainter.

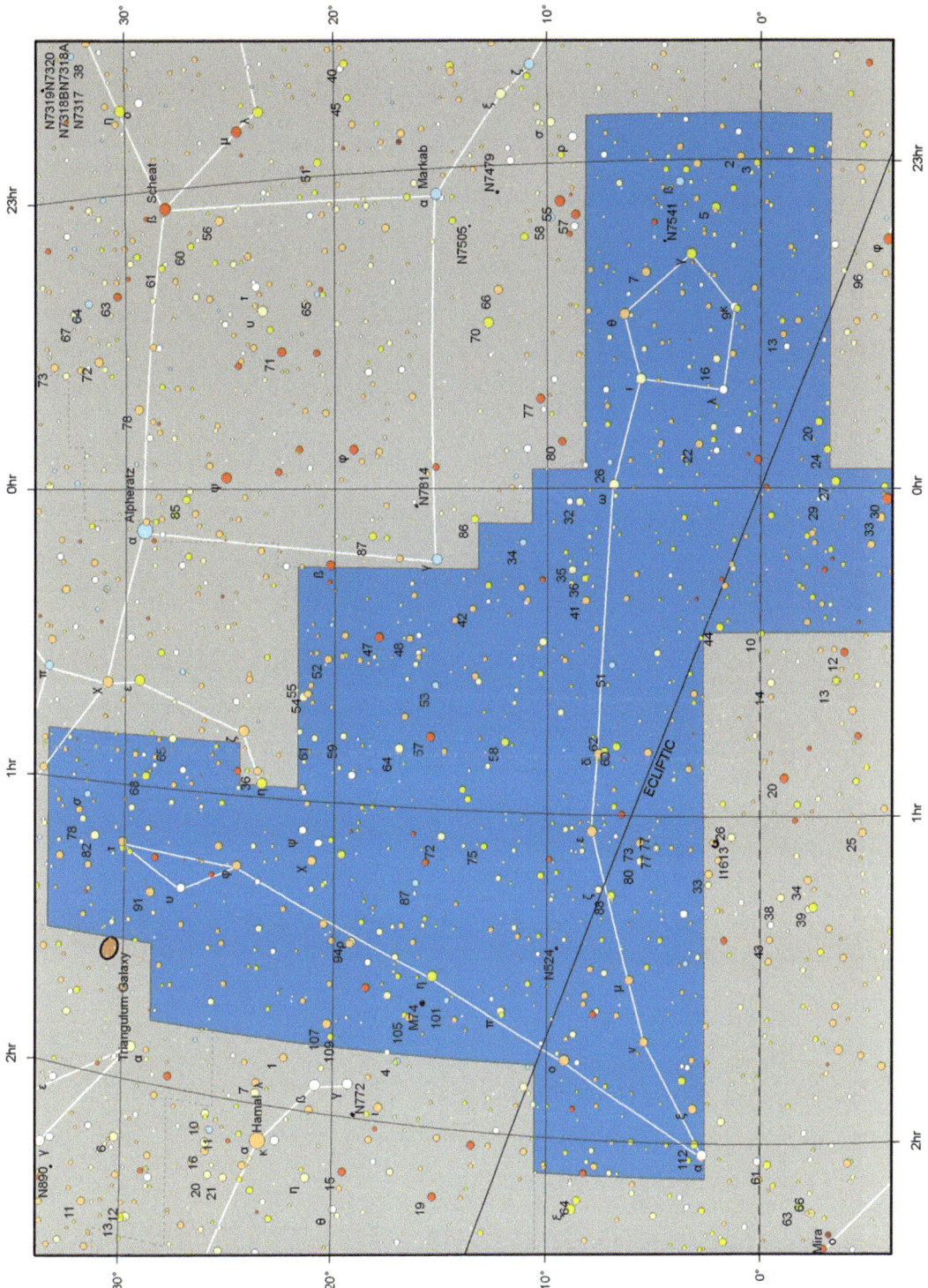

Fig. 23 M74 is a face-on spiral galaxy with a bright core, but very dim outer regions, which makes it difficult to spot, especially during Messier marathons. (Image by Grant Privett)

PISCIS AUSTRINUS

A southern hemisphere constellation that marks the southern limit of what can be observed from the UK, northern Europe and the US/Canada. It is dominated by the 1st magnitude star Fomalhaut and is pointed to by the most westerly stars of the square of Pegasus. It can be found to the south of the constellation of Aquarius.

Piscis Austrinus is most conveniently seen during the month of October.

Some very weak meteor activity may be seen to radiate from this region on or near June 27th. You will probably have to be looking out especially for it, as the peak rate is likely to be in the order of 5 meteors per hour – and those will be faint.

Historically

Another constellation with an ancient background: in this case, Babylonian. Among the legends recounted is that of the chief goddess of northern Syria and the sea, named Atargato or Derceto. Legends variously suggest she fell into a lake and was transformed into something resembling a mermaid and/or was rescued by a large fish – represented by the constellation.

Notable Stars

Double: Zeta Piscium 5.6/6.5 and 23″. White and yellow.

Double: Psi Piscium 5.3/5.5 and 30″.

Deep Sky Objects

There are no notable deep sky objects in Piscis Austrinus.

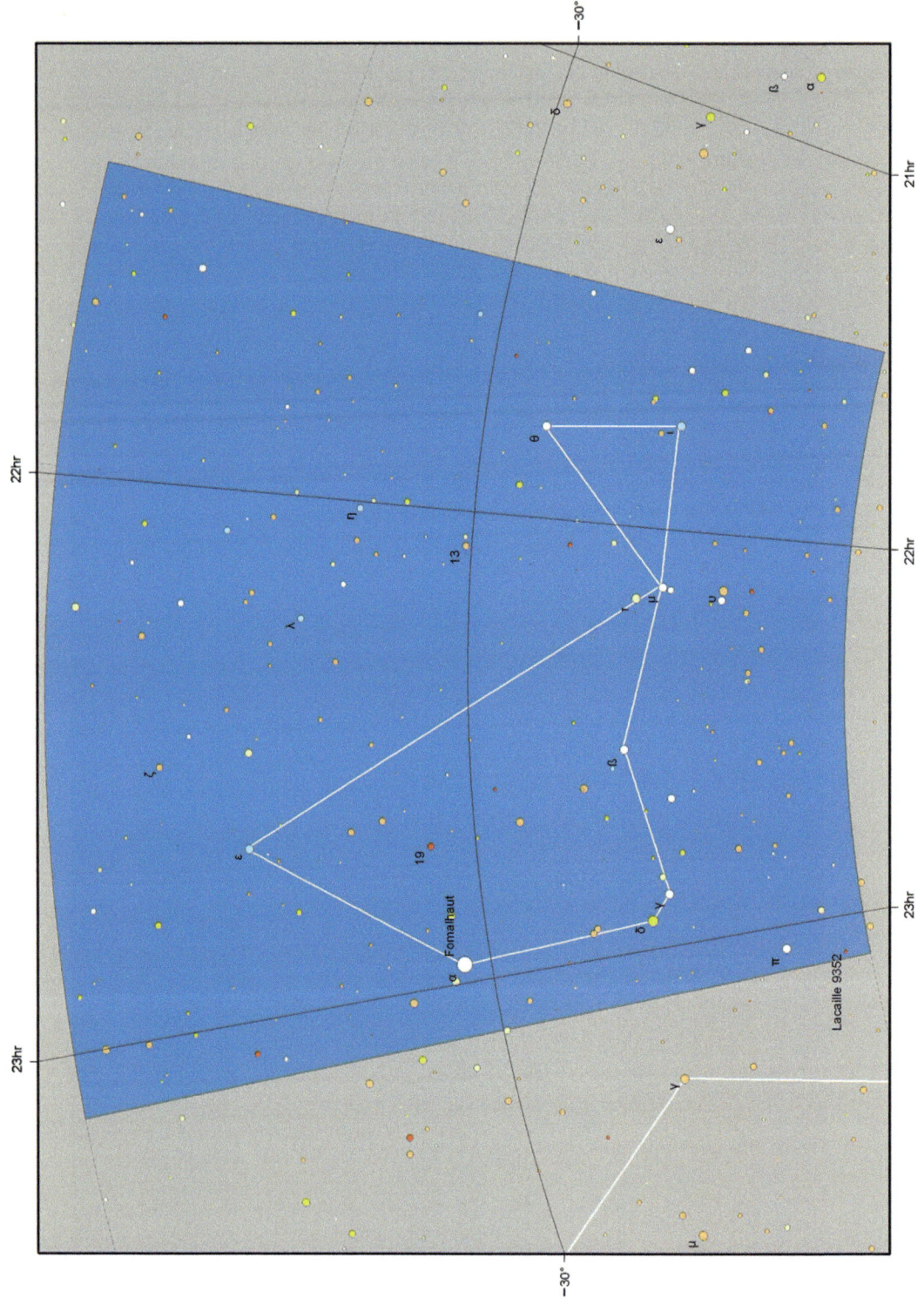

PUPPIS

A southern hemisphere constellation covering 40° of declination. It is easily found, but a little less attractive than its neighbouring constellation of Carina. It stands against a backdrop of the Milky Way and within the context of one of the most attractive regions of the night sky.

Puppis sees some minor meteor activity on 23rd April. The rate is low and the shower unpredictable and, as a consequence, the BAA and IMO would welcome your observations of the eta Puppids. Meteors may also be seen around the 6th December, when another radiant appears nearer the border with Vela.

Puppis is most easily observed during February.

Historically

The poop deck of the ship sailed by Jason and the Argonauts. This large constellation was originally part of the constellation Argo Navis, before it was broken up into the more manageable chunks of Carina (keel), Vela (sails) and Puppis.

Notable Stars

Double: Kappa Puppis 4.5/4.7 and 9.9″. White stars.

Deep Sky

M46: A bright but quite densely packed cluster hovering on the edge of naked eye visibility. Particularly notable for the foreground 10th magnitude ring-like planetary nebula NGC2438 that is in the same field of view.

M47: An even brighter – 4th magnitude – but sparse cluster just 1° away from M46. It's quite large, at nearly 30 arc minutes across. Its brightest star is magnitude 5.7. In a low power scope the two clusters can just be seen simultaneously in the eyepiece. Well seen with binoculars.

M93: Another open cluster. This time of 6th magnitude and a bit smaller, at 20 arc minutes across. Visible but unresolved in 40mm binoculars. Contains an 8th magnitude orange star and can appear triangular – depends on aperture. Some see it as butterfly shaped.

NGC2451: A bright cluster with a yellow 3rd magnitude giant included. Attractive at low powers when all its 50 stars can be seen. An interesting comparison with the nearby NGC2477.

NGC2477: A large and loose cluster a degree across, involving more than 100 faint stars. Fifth magnitude.

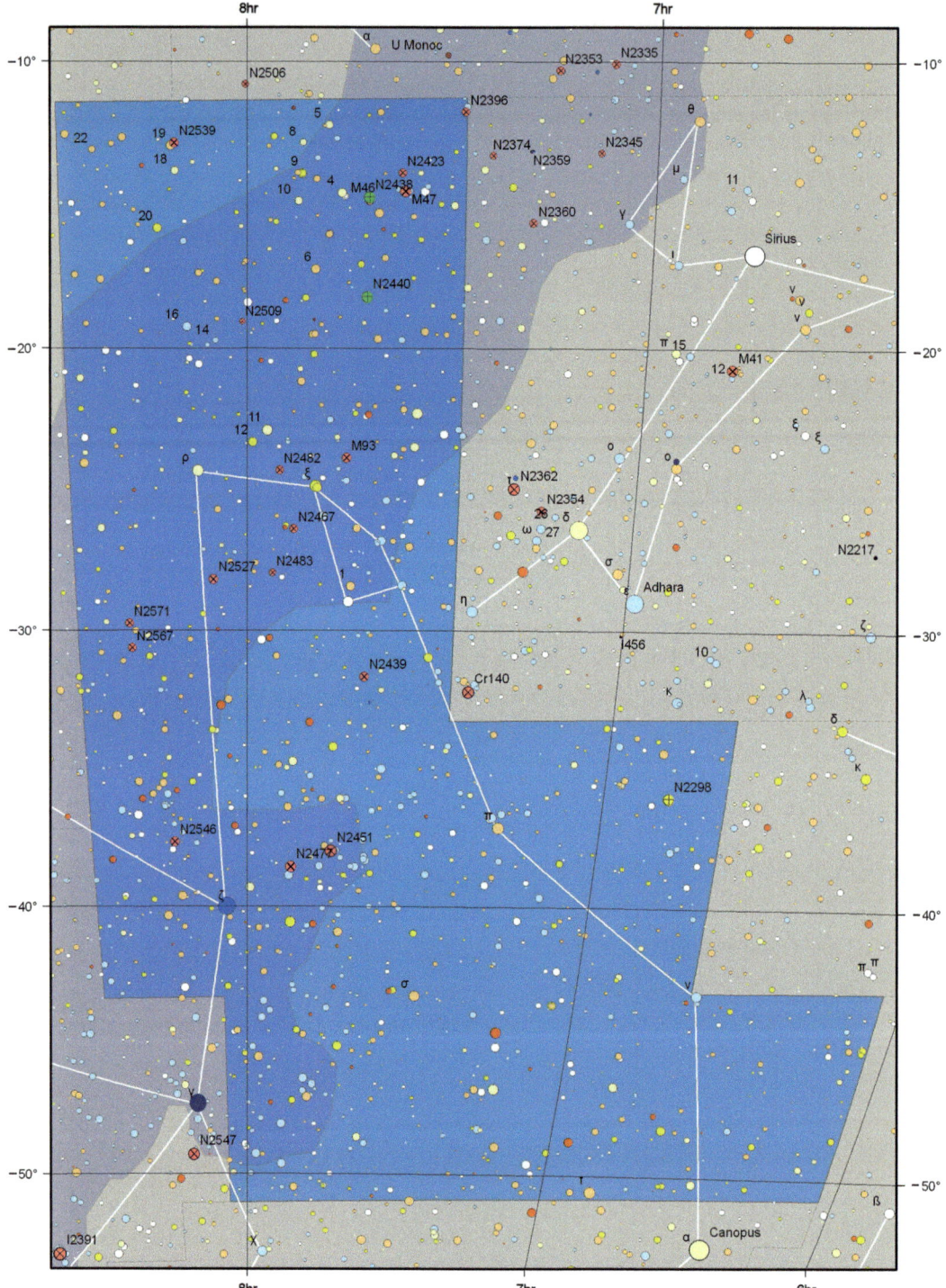

Fig. 24 The open cluster M46 and the planetary nebula NGC248 appear in the same field of view in Puppis (Image by Greg Parker)

PYXIS

A southern hemisphere constellation consisting of three 3rd and 4th magnitude stars and a handful of fainter members. Not an awe-inspiring sight and easily overlooked, given the numerous bright stars nearby and the Milky Way background.

Look for it around March.

Historically

Another of de Lacaille's inventions. It's a small, dim, constellation representing a ships compass – hence his Latin name for it of Pixis Nautica. It's very sensibly located not far from what was Argo Navis.

Notable Stars

Double: Eta Pyxidis 5.5/9.7 and 18″. White stars.

Double: Kappa Pyxidis 5.0/10.0 and 2.1″. Brighter component is a deep red carbon star.

Variable: T Pyxis is a recurrent nebula that can rise from obscurity to 7th magnitude. The intervals between the bright interludes is measured in years and the most recent event occurred in 2011, but the system is highly unstable and may one day erupt into a supernova explosion.

Deep Sky Objects

NGC2818: A patch of nebulosity 9 arc minutes across and of 8th magnitude. However, it's not especially obvious but is easy to find because of the associated open cluster NGC2818A. Look for 20–30 stars. The nebula has been imaged using the HST and is spectacular, but it may be nothing to do with the cluster.

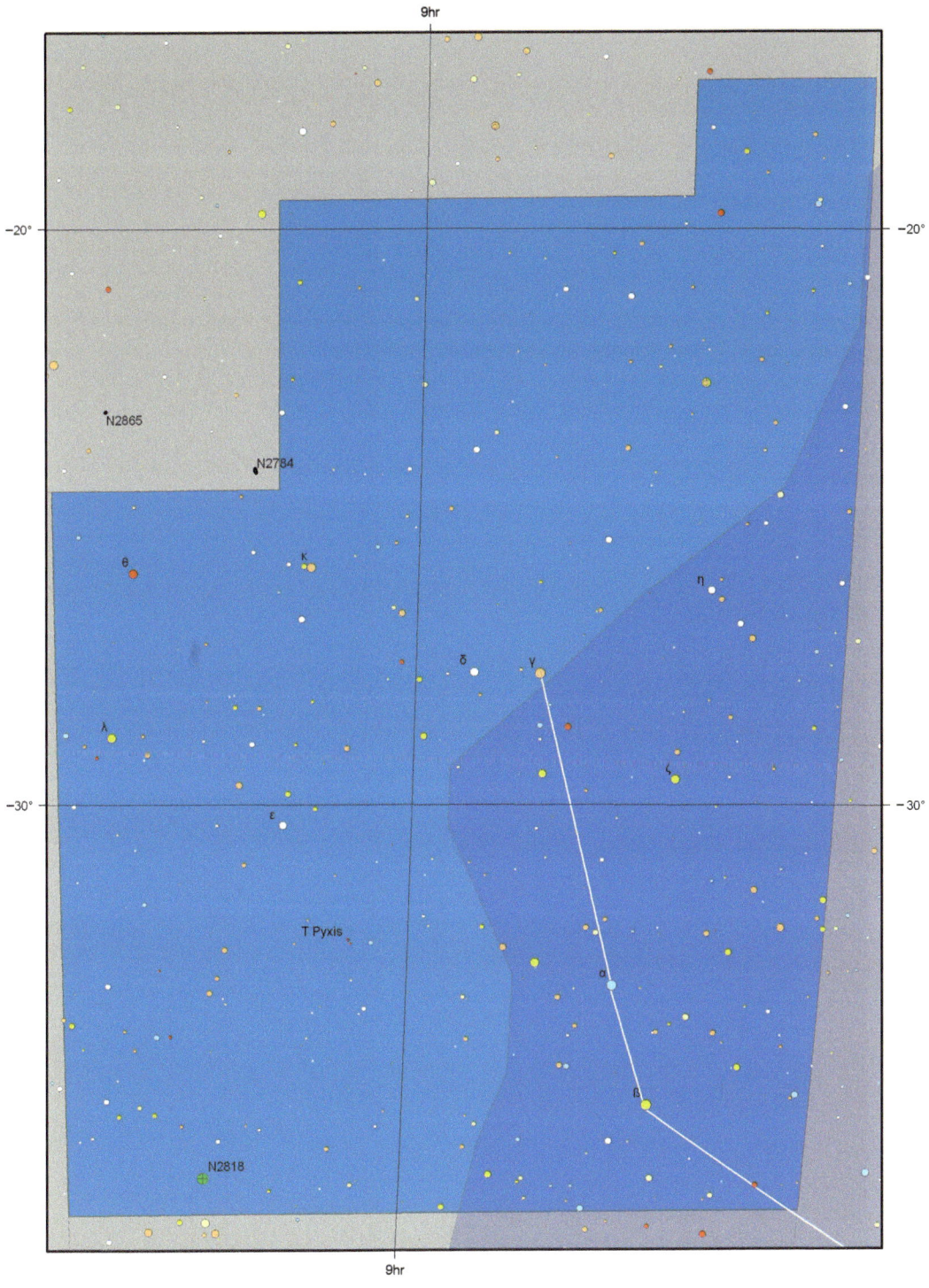

RETICULUM

A southern hemisphere constellation of 3rd and 4th magnitude stars. It isn't impressive and is best located from urban areas by examining the sky between the brilliant Canopus and impressive Achernar.

Look for it during southern hemisphere summer.

Historically

The constellation was one of those invented by de Lacaille. With this small rhombus of not particularly bright stars, he sought to immortalize the reticle eyepiece he employed when he was measuring star positions. The constellation appeared as le Réticule Rhomboide in his catalogue, so he had in part adopted the name given to this grouping of stars by the celestial globe maker Isaac Habrecht II in 1621.

Reticulum is the Latin for a small net – presumably that formed by the intersection of the two perpendicular pairs of lines seen in reticle eyepieces.

Notable Stars

Double: Zeta Reticuli 5.2/5.5 and 310″. These yellow stars should be separable with the naked eye. They are very sunlike stars.

Double: Theta Reticuli 5.9/8.0 and 4″. Blue stars.

Deep Sky Objects

There are no deep sky objects of note in Reticulum.

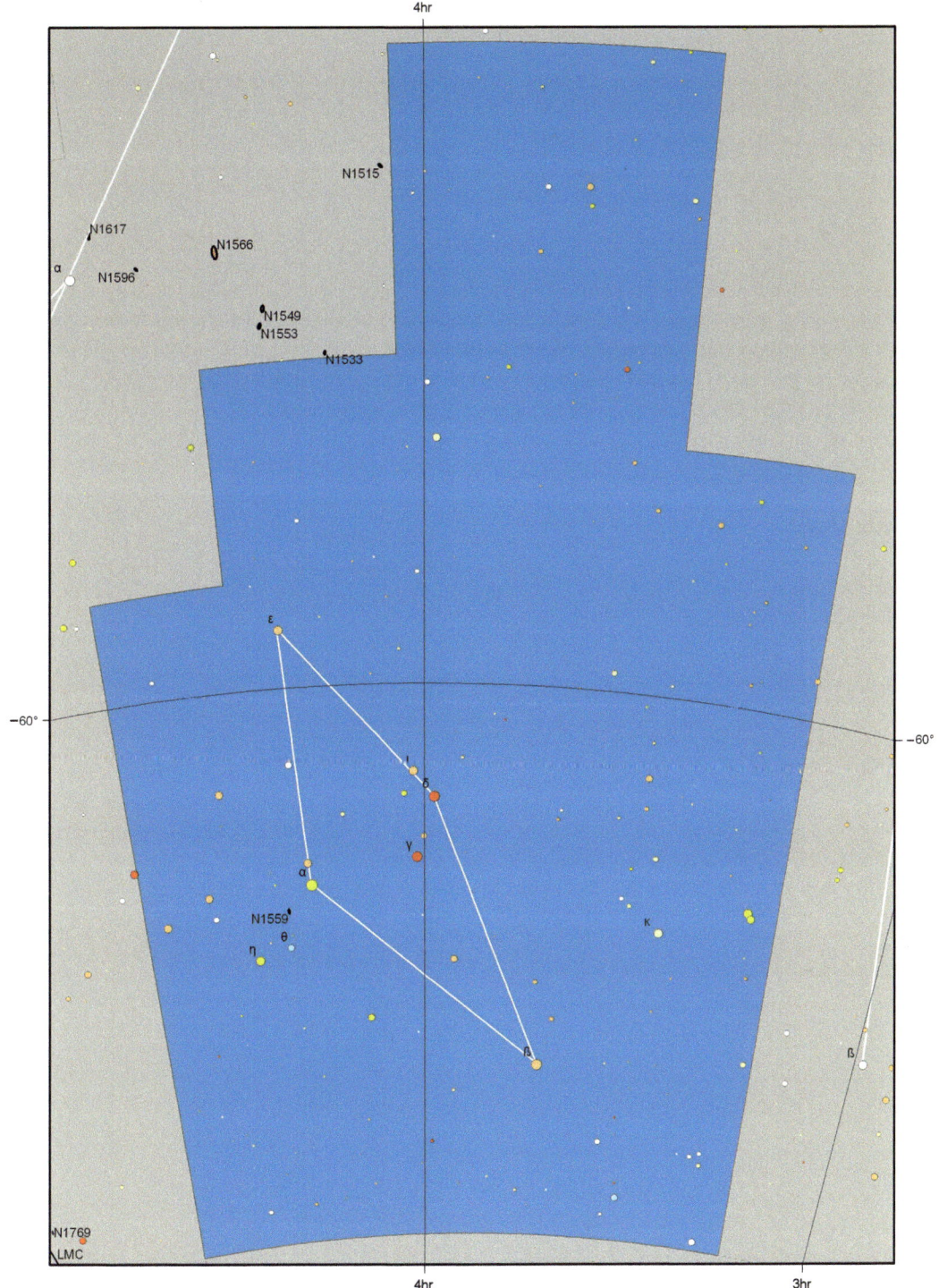

SAGITTA

Despite being quite small, Sagitta's memorable shape makes it a very familiar sight during August. It lies within the northern Milky Way, close to the Great Rift that runs down through Cygnus. It is readily located by looking north from Altair.

Historically

Sagitta is a small, but quite distinct constellation representing, and looks rather like, an arrow. Beyond that things are less clear. Whose arrow? Why fired? At whom? Prometheus, Hercules and Zeus feature large in the possible histories recounted. The Prometheus legend is particularly nasty as it involves an eagle eating his liver every day. Fortunately, or perhaps not, it regrew overnight which just perpetuated the torment of Prometheus. The arrow was used to slay the eagle, which also appears in the sky as Aquila

Notable Stars

Double: Epsilon Sagittae 5.5/7.5 and 89″. Blue and yellow.

Double: Delta Sagittae 5.5/9.0 and 8.4″. White.

Double: Theta Sagittae. 6.5/8.5 and 12″. Yellow and blue.

Deep Sky Objects

M71: A globular cluster beautifully located for easy finding, not far from Gamma Sagittae. It's a very loose affair and was for a while confused with open clusters. It's 6th magnitude and 6 arc minutes across. It's seen against a Milky Way background and is unlikely to be resolved in any aperture less than 200mm.

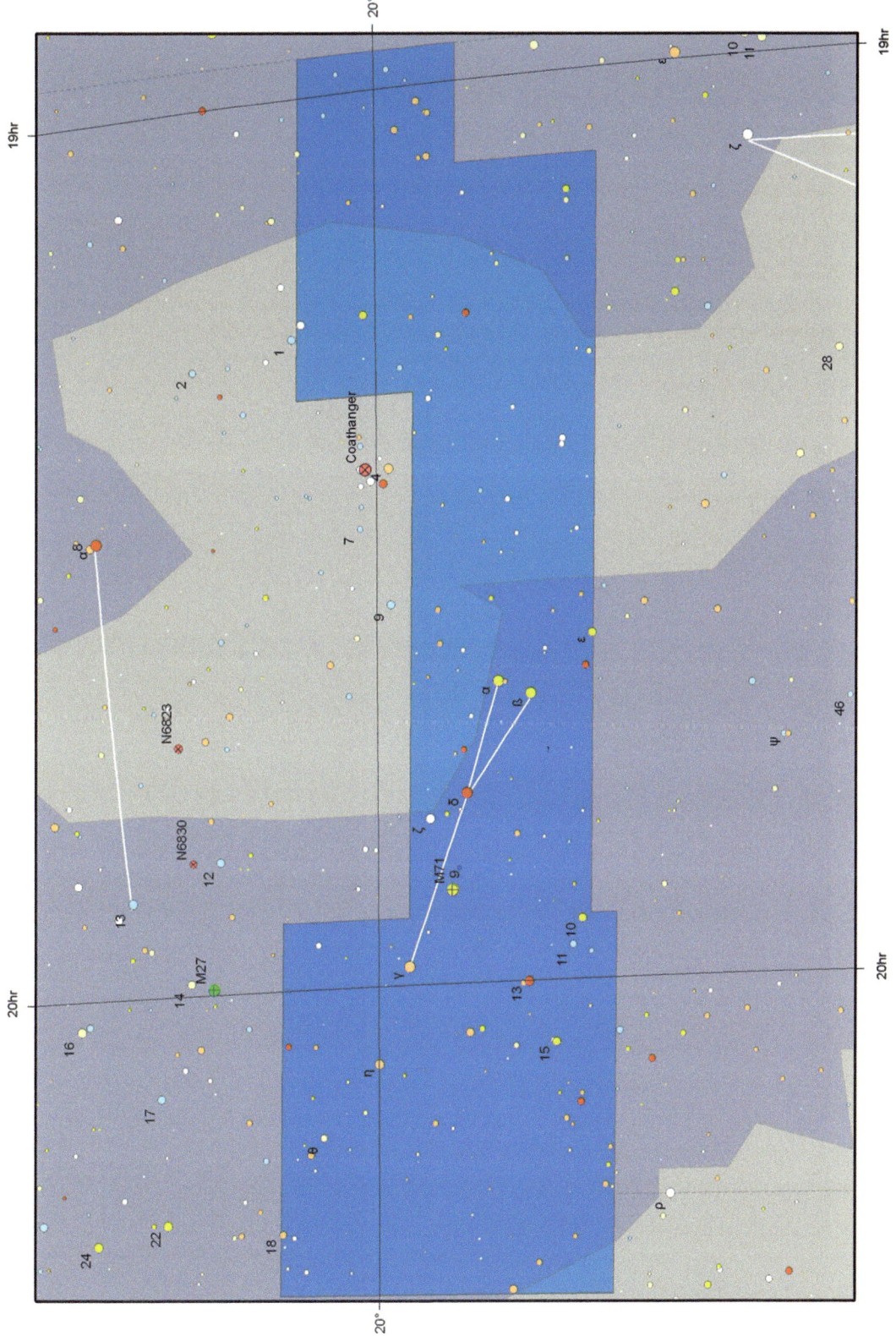

SAGITTARIUS

A southern hemisphere constellation. While its stars are not quite as bright as those of nearby Scorpius, Sagittarius does also have a very distinctive shape. It looks very much like a teapot – easy for the first time observer to pick out.

The ecliptic passes through the constellation and when we are looking at Sagittarius, we are looking towards the core of our galaxy. This ensures there is a plethora of great deep sky objects to observe. It doesn't get better than this.

Look for Sagittarius to be best placed during August.

Historically

To the Greeks, Sagittarius represented a Centaur drawing a bow and preparing to fire it. The Babylonians had other ideas, with the constellation being identified as Nergal, a somewhat mixed up creature with wings and two heads, which nonetheless fired a bow. It's not clear which head handled aiming the bow.

Nergal was a pretty busy character as he ruled the hot summer and also the netherworld where the dead were said to reside. Possibly not a good combination?

Notable Stars

Double: Beta Sagittarii 4.0/8.0 and 28″. Blue and white stars.

Double: Eta Sagittarii 3.0/9.0 and 3.6″. Tough to split the fainter component from the red primary.

Deep Sky Objects

M24: A small star cluster buried within the Sagittarius Star cloud. A particularly bright and impressive portion of the Milky Way that stands out from its surroundings. Seen from a dark location south of 30N, it's a spectacular object. 9,400 light years away.

M8: The Lagoon Nebula. It's large; it's bright. In images, it's over a degree wide. It is the best nebula in Sagittarius, and plainly visible to the naked eye from locations where it is high in the sky. Not to be missed.

M20: The Trifid Nebula is a mixture of reflection and emission nebula. Ninth magnitude and 20 arc minutes across, it is trifurcated by dark nebulae.

M17: The Omega Nebula is visible even in a 50mm aperture. An open cluster lies nearby. Two main portions of gas separated by a darker region between.

M22: A superb globular cluster of 5th magnitude, which is markedly elliptical.

M55: A seventh magnitude globular cluster readily accessible in binoculars.

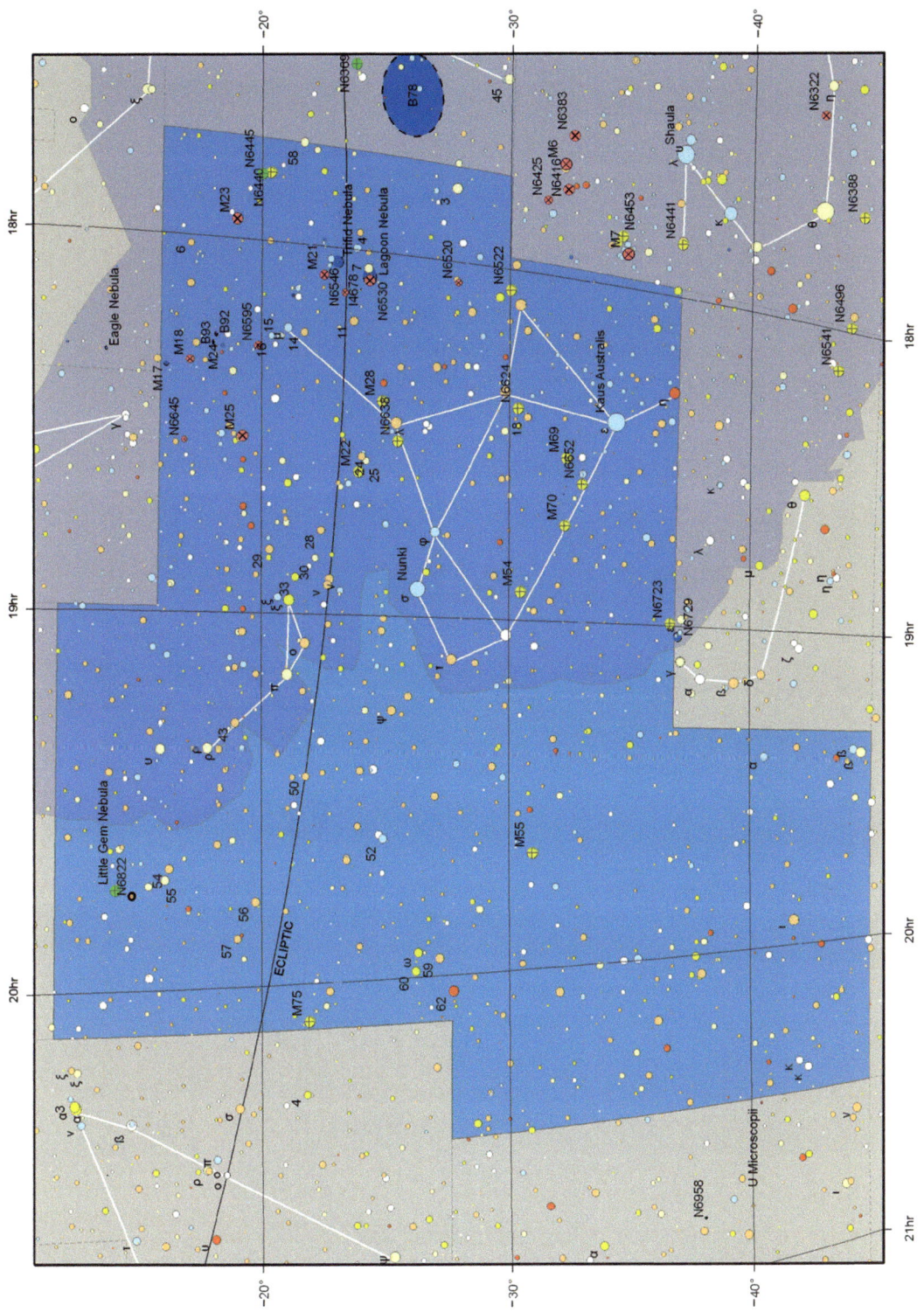

Fig. 25 M17 is a gas cloud known variously as the Omega Nebula, Swan Nebula or Horseshoe nebula. Well worth seeking out with a wide field telescope (Image by Bill Snyder)

SCORPIUS

One of the most spectacular southern hemisphere constellations, it boasts a bright and distinctive out-line. Scorpius sits in one of the richest portions of the Milky Way and provides wonderful views for binocular and telescope users.

It is dominated by the 1st magnitude red giant star Antares and, due to the ecliptic passing through the constellation, you may sometimes get to contrast its color with that of the brighter planets, notably the ruddy Mars. Actually, it varies in brightness, between magnitude 0.9 and 1.2.

Scorpius is best placed for inspection during July.

Historically

Scorpius, known to astrologers as Scorpio, is an ancient Sumerian constellation that really does look like its namesake. Its grand form dominates the southern horizon during the northern hemisphere sum-mer in a similar manner to that of Orion which dominates the northern hemisphere winter sky. The legend has it that they are separated on the sky because Scorpius stung Orion to death after Orion either attempted to ravish Artemis, or became overly boastful.

Notable Stars

Double: Xi Scorpii 5.0/5.5/7.0 52″. White stars.

Double: Beta Scorpii 2.5/5.0 13.6″. Blue and white stars. You are only seeing part of a very complex system of six or more components.

Double: Nu Scorpii 4.3/6.5 and 41″. The dimmer star is itself a close double.

Double: Alpha Scorpii 1.0/6.5 2.6″. Often perceived as red and green, due to contrast and brightness differences. Tough in small scopes, but worth a try.

Deep Sky Objects

M4: A magnificent globular cluster. Easily found by offsetting from Antares – within the same binocular field. Large and 7th magnitude, it resolves well in a 150mm aperture. Contains attractive eye-catching lines of stars but is not strongly condensed.

M80: A 7th magnitude globular cluster that is very highly condensed. The brightest members are 14th magnitude so a 200mm + aperture will be needed for resolution. Much smaller than the brighter M4, it is 33,000 light years away.

M7: A superb, coarse, naked eye (3rd magnitude) open cluster near the Scorpions sting. Quite large at over a degree across, it was mentioned by Ptolemy.

M6: An open cluster well suited to smaller apertures. Twenty arc minutes across and of 4th magnitude. Some see a butterfly pattern in the star field of 100 stars.

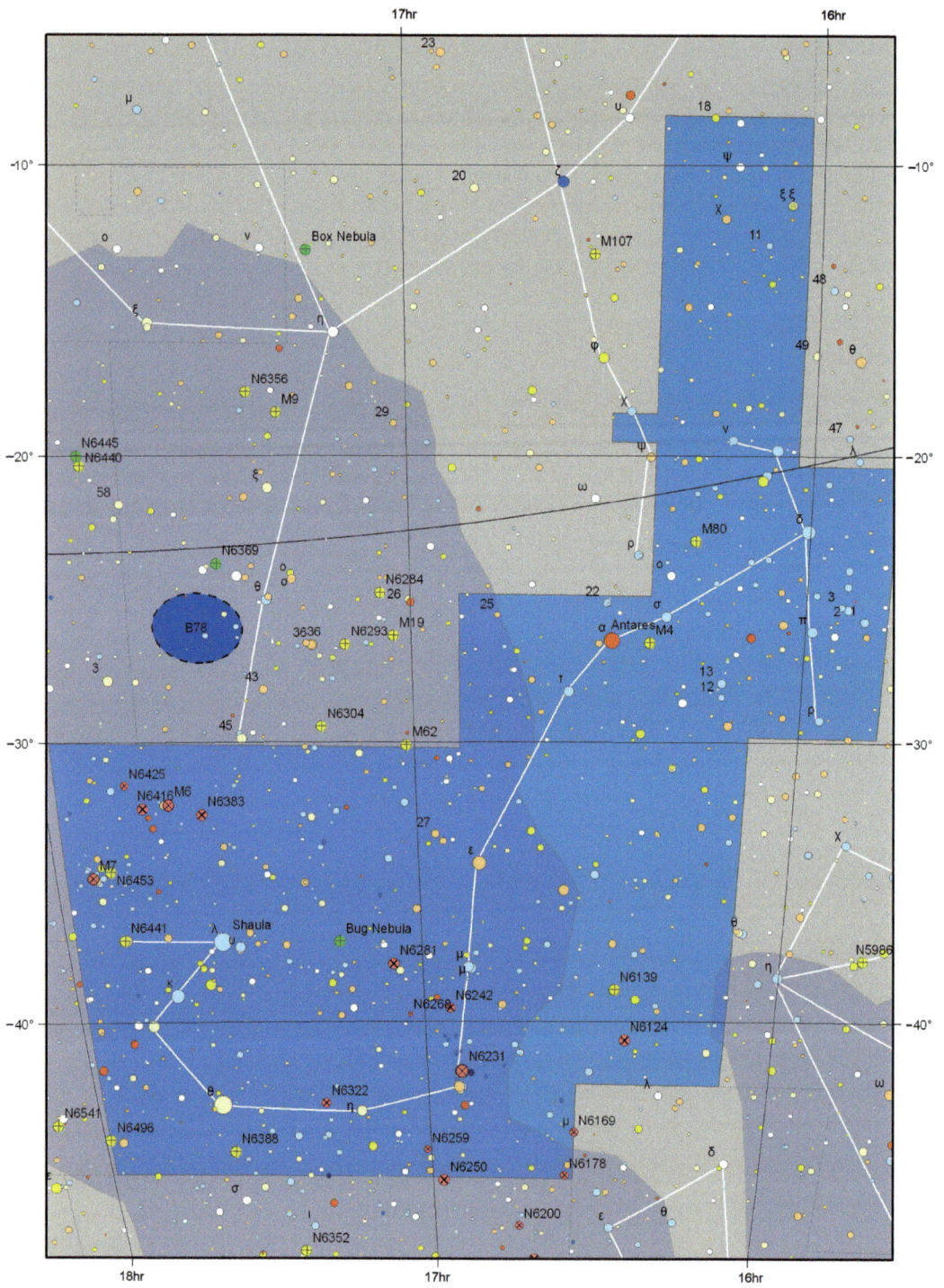

SCULPTOR

An unimpressive southern constellation.

Sculptor has no stars brighter than 3rd magnitude and is notable only for the presence of the South Galactic Pole within its boundary. So when you are looking at Sculptor you are looking perpendicular to the plane of our galaxy and outward into intergalactic space – which is why, when viewed in telescopes, there are surprisingly few background stars.

Sculptor is also host to a cluster of galaxies only slightly further away from us than our local group of galaxies.

Historically

Yet another de Lacaille constellation. It's the second that mentions sculpting. The first was Caelum – which was originally Caelum Sculptorium – while the modern Sculptor is a shortened version of Apparatus Sculptoris – the Sculptor's studio. The original star chart shows a three-legged table supporting a bust that's being worked on using chisels and a hammer.

Notable Stars

Double: Epsilon Sculptoris 5.4/8.6 and 4.7″. White and yellow color

Double: Delta Sculptoris 4.6/9.5 and 75″.

Deep Sky Objects

NGC253: The brightest galaxy in the Sculptor galaxy cluster, this was one of Caroline Herschel's discoveries. A large and bright inclined spiral galaxy undergoing a burst of active star formation. Eighth magnitude and nearly 20 arc minutes long by 6, it may even be well seen in your finderscope.

NGC55: A large (30×5 arc minutes) edge on irregular galaxy that appears mottled in 200mm apertures. It may be gravitationally interacting with another galaxy of the Sculptor galaxy cluster, NGC300. Seventh magnitude. May be difficult to see in apertures less than 150mm.

NGC288: A relatively loose globular cluster of 9th magnitude, which spans 9 arc minutes.

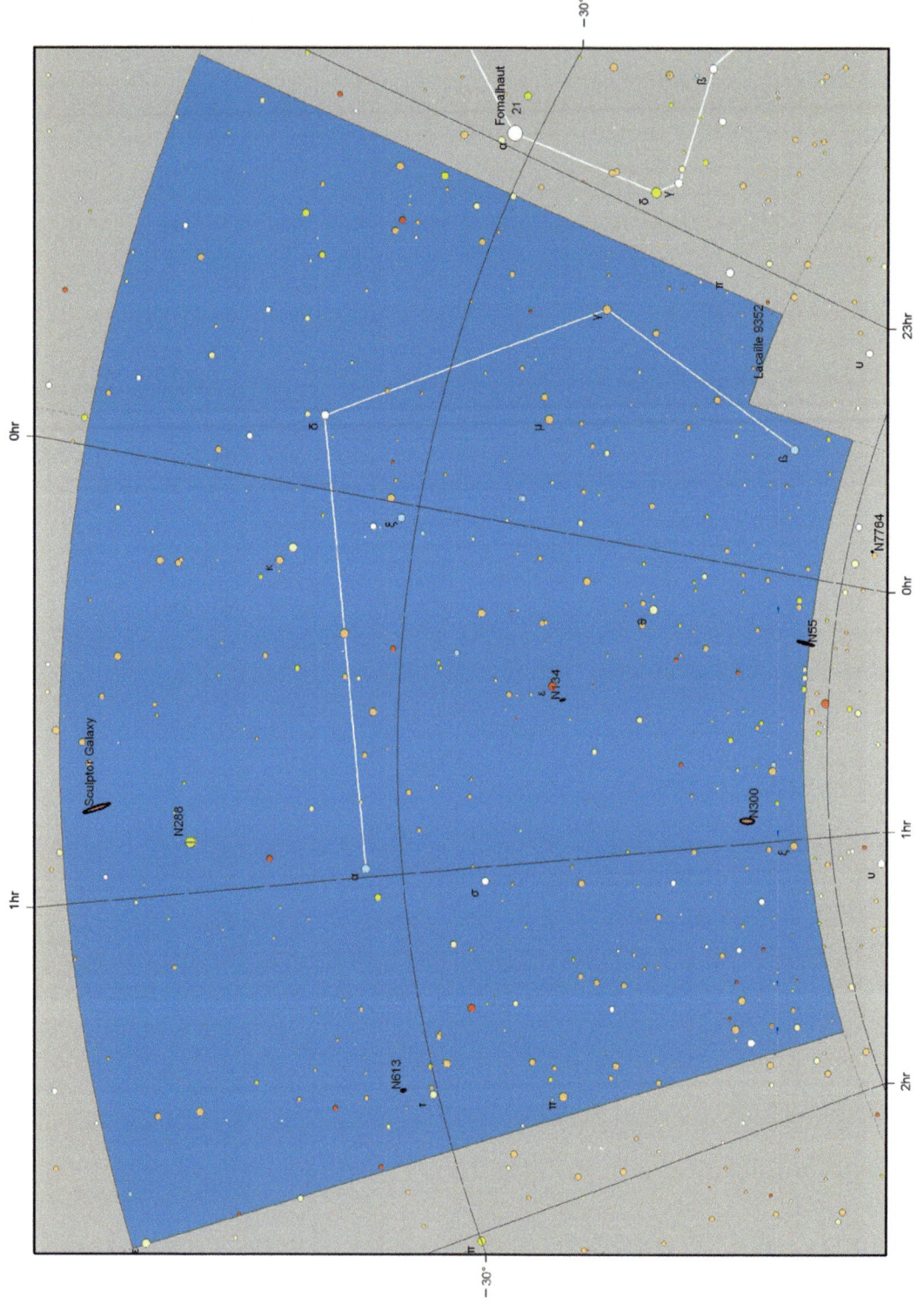

SCUTUM

This small unassuming constellation lies just south of the celestial equator and is embedded deep in one of the most impressive portions of the Milky Way – an area rich in detail and feature that rewards any imaging.

Scutum is very well worth a visit and provides splendidly rich and attractive fields when seen in binoculars.

Look for it to be best placed for observation during August.

Historically

A constellation introduced on charts created by Johannes Hevelius in 1684. Being a good and loyal Polish citizen, he created a constellation to immortalize King John III of Poland, who was a patron and major source of funding for Hevelius.

It depicts a shield, so the constellation initially rejoiced in the name Scutum Sobiescianum, as King John's family name was Sobieski. This astute bit of flattery certainly didn't do Hevelius any harm and, despite it, the constellation has remained until the present day (Whereas several other blatant attempts to butter up patrons were rejected by the astronomical community over the years.)

Notable Stars

Double: Delta Scuti 4.5/9.4 and 52″.

Deep Sky Objects

M26: A sparse open cluster spanning a quarter of a degree. The brightest member is 10th magnitude. May be difficult to spot against the local, rich, star background.

M11: By way of contrast, The Wild Duck Cluster is a rich, bright and very obvious open cluster. It contains more than 1,000 stars – only 200 or so are likely to be seen in smaller instruments – tightly packed into a third of a degree of sky. The brightest member is 8th magnitude, but 9th magnitude is more typical. Highly concentrated. A must-see object well worth tracking down.

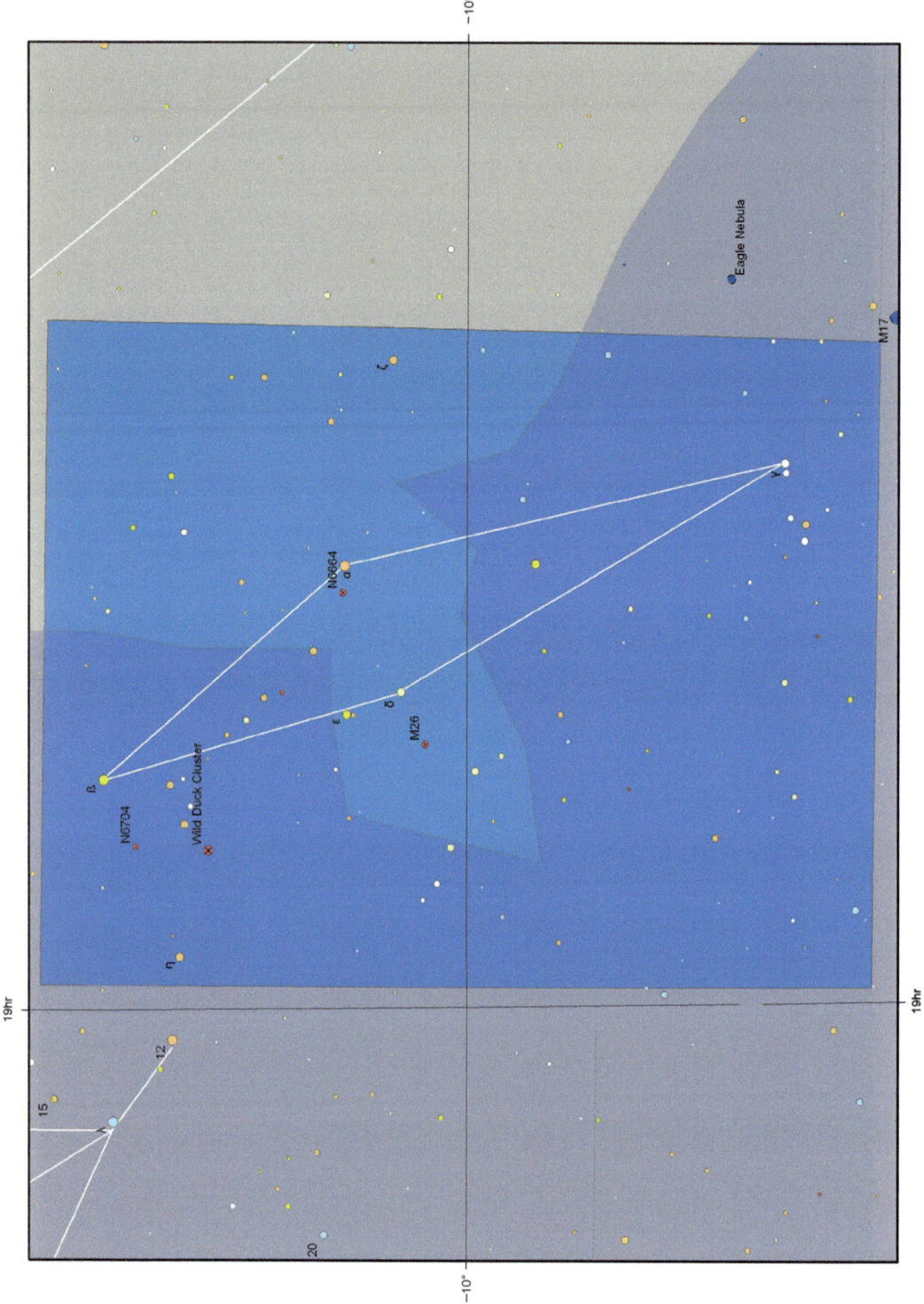

SERPENS CAUDA AND SERPENS CAPUT

A unique constellation, Serpens consists of two parts, representing the head and tail of a snake that is being grasped by Ophiuchus. The constellation bestrides the celestial equator. The ecliptic also passes through the southern part of Ophiuchus between Scorpius and Sagittarius – something conveniently forgotten by astrologers.

Serpens lies just to the west of one of the most impressive portions of the Milky Way.

Historically

One of 48 constellations described by Ptolemy in his *Almagest*, its stars depict a serpent/snake held by, or wrapped round, the figure of Ophiuchus – believed by the Romans to be a great healer. The makers of early charts differed on Serpens, with some representing the snake as a constellation in its own right, while some others chose to show Ophiuchus and the snake together as a single entity.

The first explicit record of the constellation is dated around 400 BC, but it probably derives from a still older source.

Notable Stars

Double: Delta Serpentis 4.2/5.2 and 3.9″. White stars.

Double: Beta Serpentis 3.5/9.0 and 31″. White stars.

Double: Theta Serpentis 4.5/5.2 and 22″. Blue-white stars.

Deep Sky Objects

M5: A fine globular cluster. From good sites it has been spotted naked eye, but it is plainly visible in binoculars and can be resolved into stars in a 200mm aperture. Ten arc minutes across.

M16: The Eagle nebula has become one of the most familiar nebulae, after the publication of the famous HST image, but it isn't easy to see. The encapsulated cluster is more obvious. The cluster is coarse and small – just 8 arc minutes across and contains members that are of 7th and 8th magnitude.

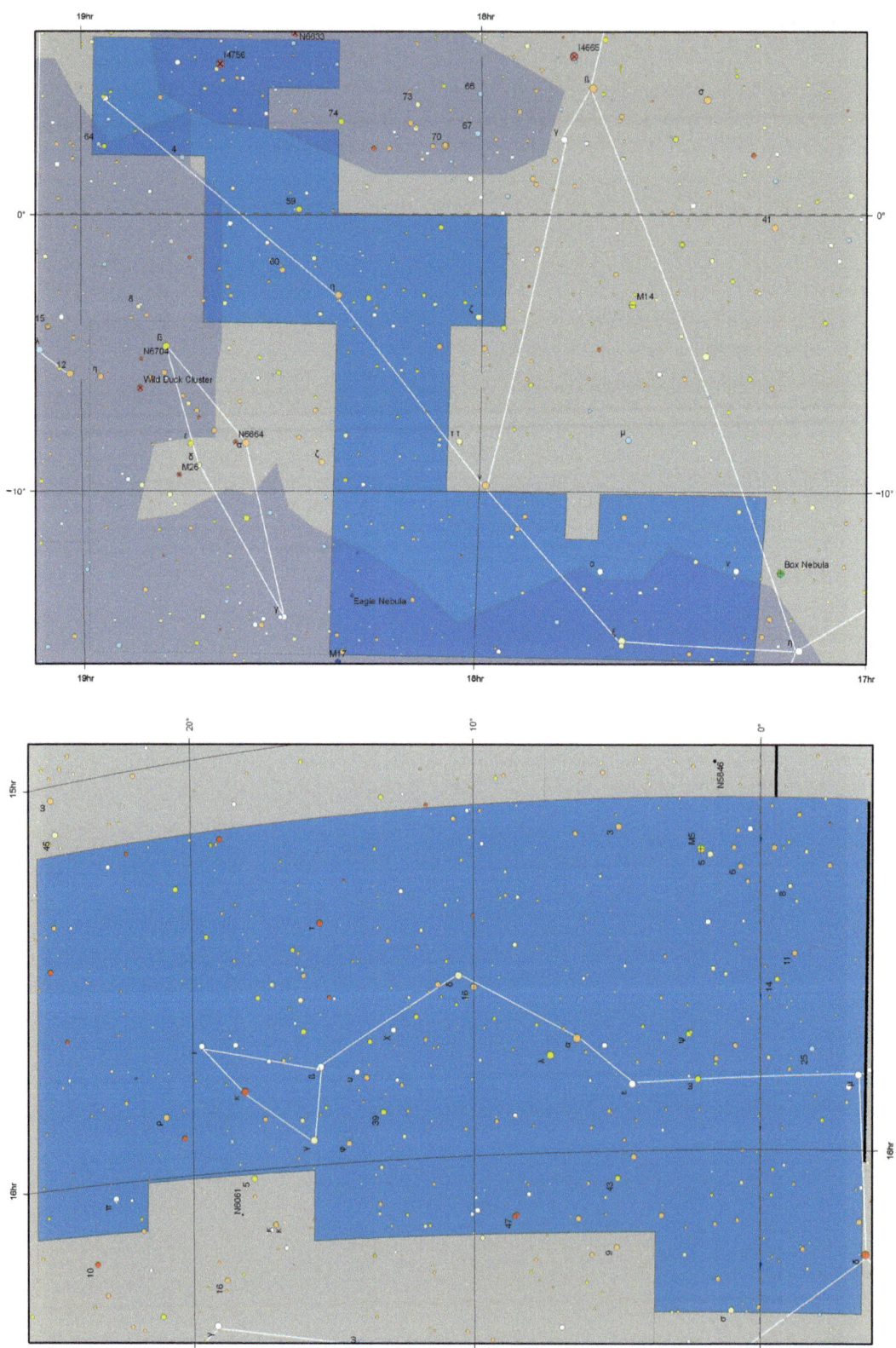

SEXTANS

A northern hemisphere constellation composed of 4th and 5th magnitude stars located just above Hydra. It lies on the Celestial Equator and is very close to the ecliptic, so the Moon will occasionally be seen within its bounds. It is best placed for observation during the month of April.

Historically

A constellation invented by Johannes Hevelius to depict an instrument, the sextant, used to measure star positions. Hevelius was an interesting man, not only Mayor of Danzig at one point, but also head of the local beer-brewing guild. He probably had a lot of friends.

Notable Stars

No double stars.

Deep Sky Objects

There are no notable deep sky objects in Sextans.

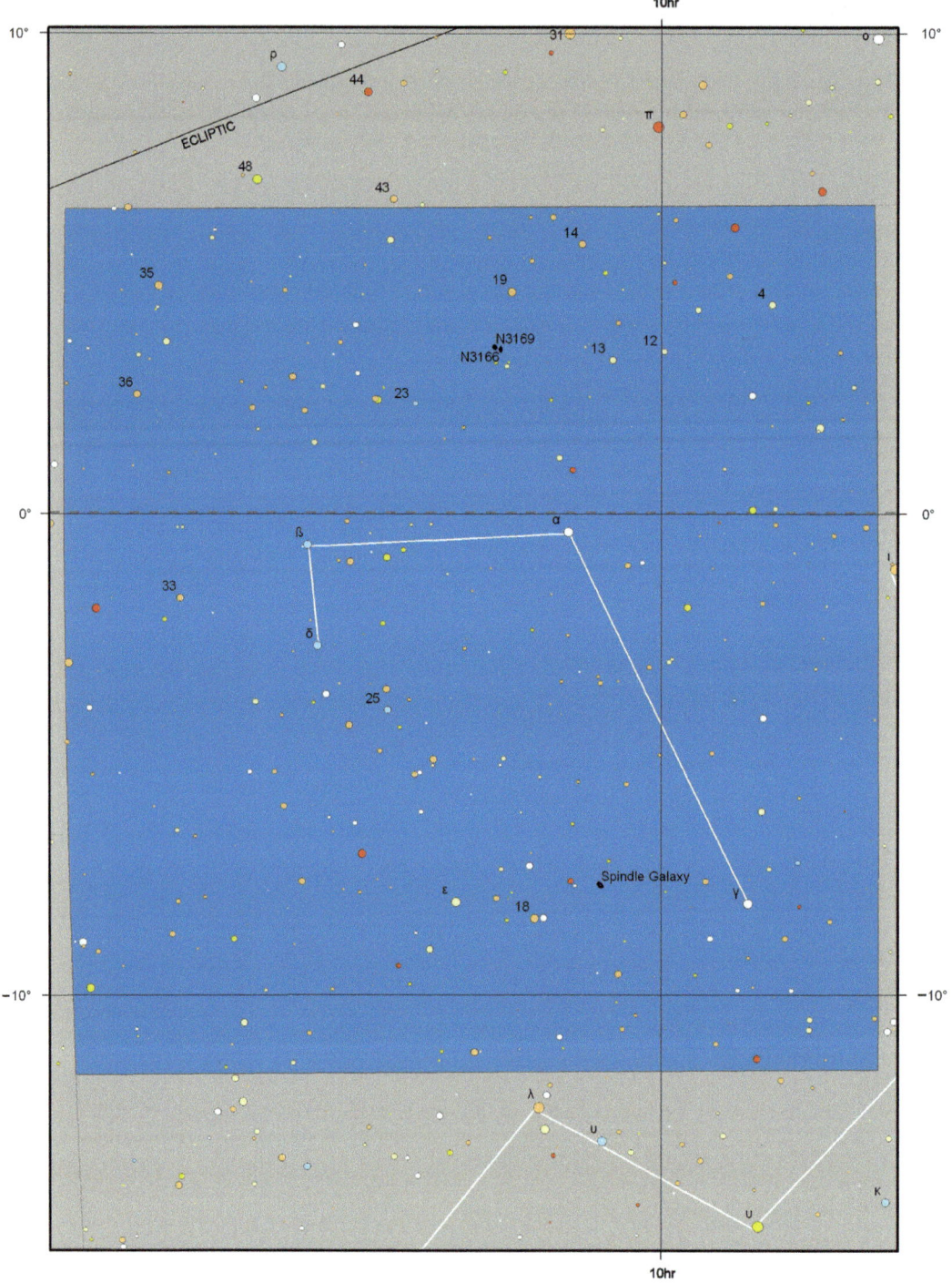

TAURUS

An impressive northern hemisphere constellation dominated by the open cluster, The Hyades, which provides a distinctive "v" shape that is a background for the brilliant orange star Aldebaran. The eastern end of the constellation is on the outskirts of the Milky Way, making for interesting sweeping. The ecliptic runs through the constellation, so it will often play host to a planet or the Moon.

Two minor meteor showers – more like drizzle in these cases – run from October through to December. The rate isn't high – 10, or so meteors an hour – but there are peaks in October and November. They arise from a radiant that sprawls across the Taurus/Cetus border.

Historically

Greek legends associated with Taurus, the bull, revolve, as is so often the case, around Zeus. Taurus is said to depict the heifer into which Zeus transformed himself when seeking to abduct the Phoenician maiden Europa. He became a particularly pretty and docile heifer with a silver coat that Europa encountered on a beach. So taken was she with its placid nature that she rode on its back and was then was carried away to Crete by Zeus. It's amazing anyone ever thought Zeus worthy of worship.

Notable Stars

Double: Chi Tauri 5.5/7.5 19.4″. Blue and white.

Double: Phi Tauri 5.0/8.4 and 52″. Yellow and blue stars.

Double: Theta Tauri 3.5/4.0 and 335″. Yellow and white.

Double: Tau Tauri 4.0/8.0 and 63″. Blue and white.

Variable: R Tauri is a Mira type star that changes between magnitudes 7 and 15 in around 320 days. It has an associated cloud of dim nebula, which has been found to be one of the few "variable nebulae" in the sky.

Deep Sky Objects

M45: The Pleiades. One of the finest clusters in the sky. A stripped down truncated Plough made of white-blue stars. A wonderful sight in binoculars or a telescope – the wider the field the better. Naked eye from a suburban location, six members will normally be seen. Hundreds of stars are actually present and drawing it becomes tough if the aperture exceeds 114mm.

Hyades: A large "V" of 200+ stars near Aldeberan, but actually twice as far away. Many members of the cluster are visible to the naked eye.

M1: The Crab nebula is the brightest supernova (1054 AD) remnant in the sky. An elongated, and expanding, patch of debris of 8th magnitude, containing a pulsar. It can be spotted with a 114mm aperture but detail will prove elusive.

NGC1746: A loose open cluster 45 arc minutes across in a nice star field.

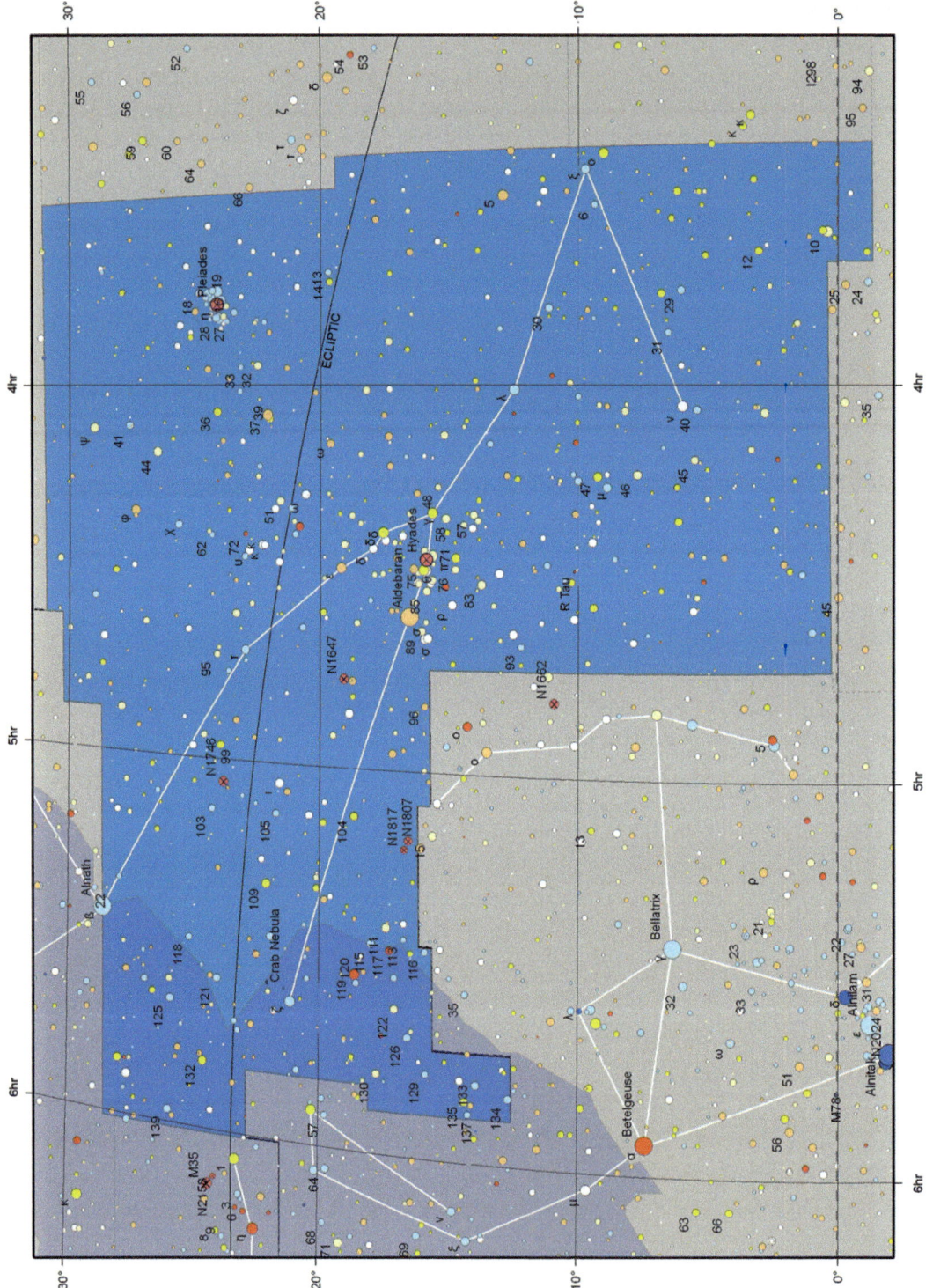

Fig. 26 The naked eye cluster, the Pleiades. A collection of several hundred bright young stars in Taurus (Image by David Ratledge)

TELESCOPIUM

A rather dull southern hemisphere constellation that bears little resemblance to its namesake. It is rather overshadowed by its gaudier neighbors Scorpius and Sagittarius and will be most easily observed when they culminate through August.

Historically

Continuing his theme of honoring scientific devices in the sky, de Lacaille created a constellation, Telescopium, to represent the refracting telescope. The original chart drawing, Lacaille's southern planisphere, shows the tube assembly hoisted using pulleys, suggesting one of the long focal length "aerial" telescopes that were popular at the Paris observatory when de Lacaille was young. These were much used by his patron, Jacques Cassini.

Notable Stars

No double stars.

Deep Sky Objects

There are no notable deep sky objects in Telescopium.

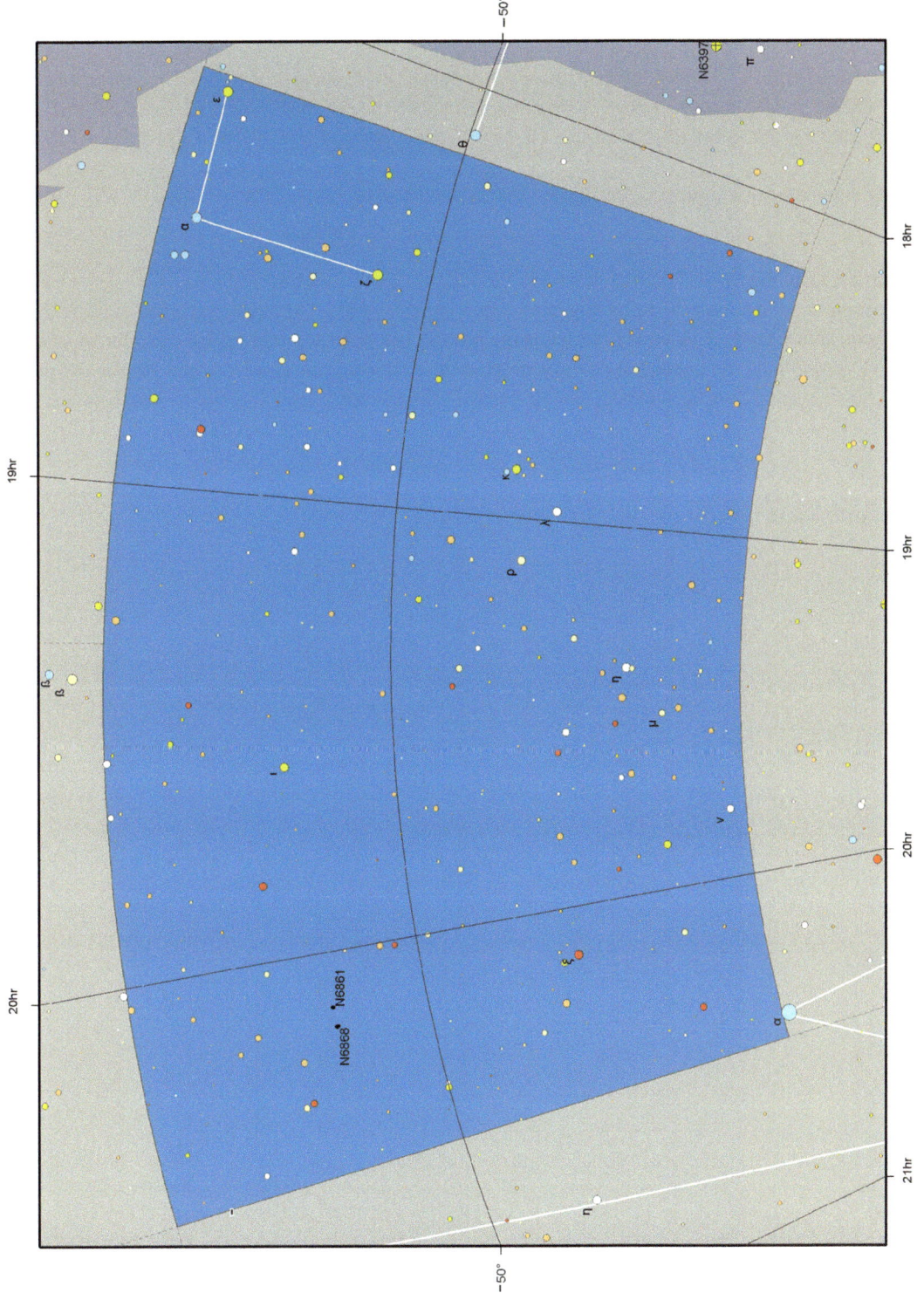

TRIANGULUM

A northern hemisphere constellation formed by three stars of 3rd and 4th magnitude. It's not as bright as its southern counterpart, but is easily located near Aries and below Andromeda. Well worth seeking out for its single bright galaxy, but beware, it can be difficult to spot under suburban skies and is impossible in the cities.

Triangulum rides highest in the skies at the end of the year.

Historically

A constellation familiar to the Greeks and Romans, its stars make up the form of an isosceles triangle (two sides have the same length) not far from Aries. The Babylonians also recognized it as a constellation but appear to have seen it as a plough shape – perhaps integrating some of the stars now within Andromeda.

Notable Stars

Double: Iota Trianguli 5.3/6.8 and 3.9″. Yellow and white.

Variable: R Trianguli is a Mira type variable star that can be watched changing over a 270 day cycle from magnitude 5 to magnitude 12 i.e. naked eye to the limit of an 80mm refractor.

Deep Sky Objects

M33: Triangulum's sole claim to deep sky fame. This face-on spiral galaxy, has an integrated magnitude of 6th magnitude, but because it is a degree across, it has a low overall surface brightness and can be very difficult to spot under light polluted skies. It is best seen for the first time via binoculars – 10×50s or 12×80s. It can be difficult to see with the naked eye but is accessible from a site with a limiting magnitude of 7.

It's a member of the local group of galaxies and images well but it can be a little disappointing in amateur scopes: especially from urban and suburban areas.

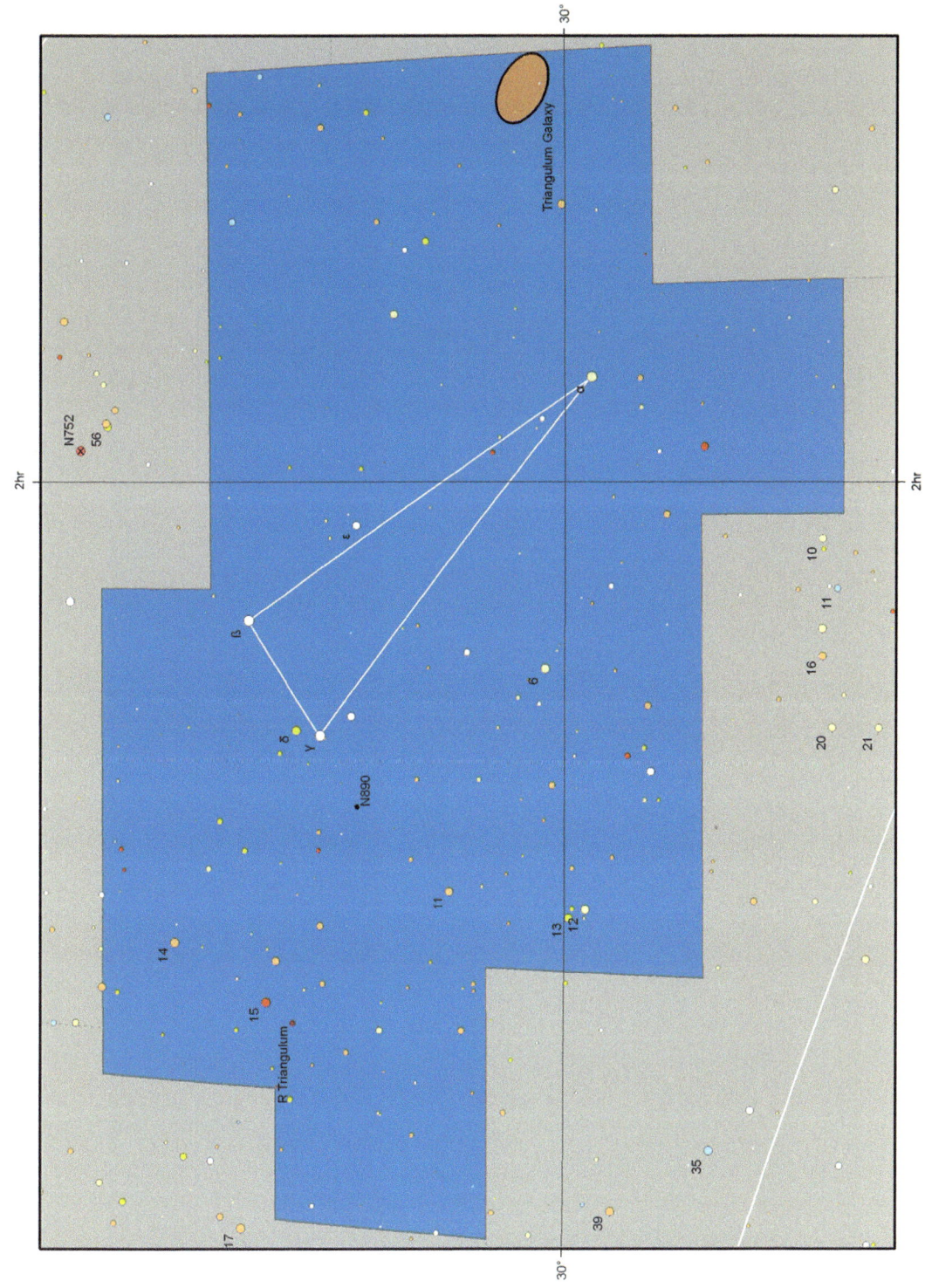

TRIANGULUM AUSTRALE

A small, unimposing southern hemisphere constellation lying not far from alpha and beta Centauri. It does at least look like its name, but any three stars would! It is best hunted down during July, but it really has very little to offer the observer: a single 1st magnitude star, two second magnitude stars and a single worthwhile cluster.

The fifth smallest constellation.

Historically

This constellation is not visible from Europe and so the Romans, Greeks and Sumerians have nothing much to say about it. Instead, it's a small constellation introduced by Plancius on his sky globe in 1589. The stars are reasonably bright, but there is little in the way of recorded associated history

Notable Stars

No double stars.

Deep Sky Objects

NGC6025: An open cluster of 7th magnitude. Fifteen arc minutes wide and containing 30 or so stars. Half the stars are visible in a 75mm aperture – some are as bright as 7th magnitude. The cluster is observable in binoculars.

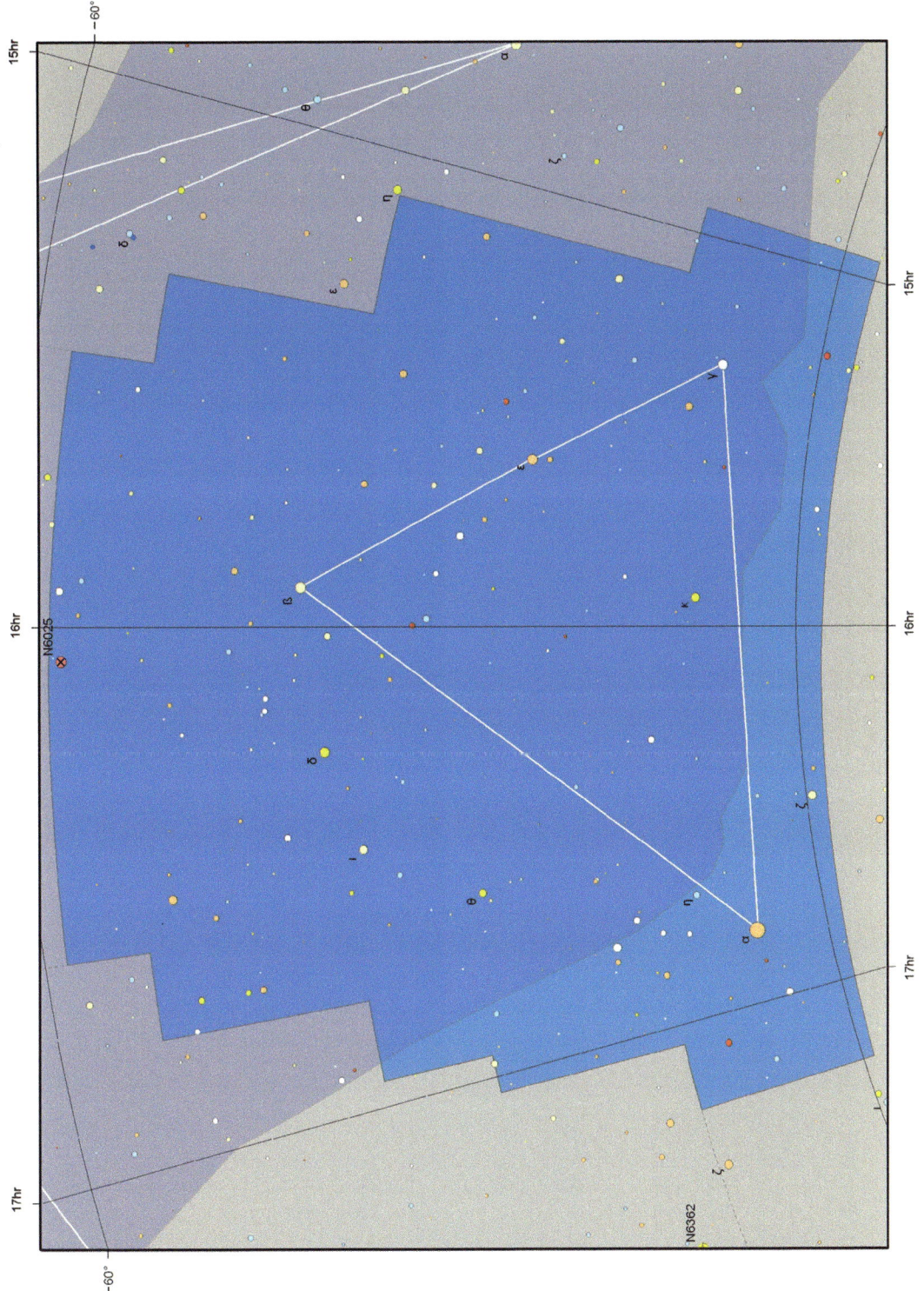

TUCANA

A southern hemisphere constellation lying in a reasonably sparsely populated portion of the sky, not far from the bright star Achernar and also the South Celestial Pole of the sky. Other than 2nd magnitude alpha Tucanae, the stars are not impressively bright but it does have several worthwhile double stars.

However, what it lacks in stars it certainly makes up for in deep sky interest. The Small Magellanic Cloud sits near its southern boundary with, close-by, a globular cluster so bright it got labelled as a star. A deep sky observers heaven.

Historically

One of the constellations that remain from the 21 invented by Plancius. It depicts a large billed bird. It's reported as a Toucan but may, originally, have been based upon a different bird. It appears in Johann Bayer's 1603 *Uranometria Omnium Asterismorum* star atlas – one of the first to depict stars encompassing the whole sky.

Notable Stars

Double: Beta Tucanae 4.2/4.5 and 27″. Blue stars. A complex system of 5 or 6 stars. The two bright components are each, in their turn, doubles.

Double: Kappa Tucanae 4.5/8.7 and 7.0″. White and red. Worth a look.

Double: Delta Tucanae 4.5/9.0 and 7″. Blue and white stars.

Sunlike star: Zeta Tucanae is sunlike, sharing many properties with our Sun.

Deep Sky Objects

NGC 292: The Small Magellanic Cloud (SMC) is a satellite galaxy of our own Milky Way. It's a dwarf galaxy boasting hints of a bar structure, suggesting a distant past as a spiral galaxy before interactions with our galaxy distorted it.

The galaxy is very plainly visible to the naked eye – to my sight, it's brighter than the Milky Way itself, and spans an area more than 5° across. A southern hemisphere must-see object.

NGC362: A 6th magnitude globular cluster hovering on the edge of the SMC. Resolvable in a 150mm aperture it has a condensed, compact core.

NGC104: The famous 47 Tucanae globular cluster. Obvious to the naked eye, and large. Looks very much like the coma of a comet. Even without the SMC this cluster would make visiting Tucana well worth the effort. Visible in binoculars and resolving well in small instruments, its stars sprawl across more than 30 arc minutes. A small open cluster, NGC121, shares the same field of view, but may go unnoticed in such fine company.

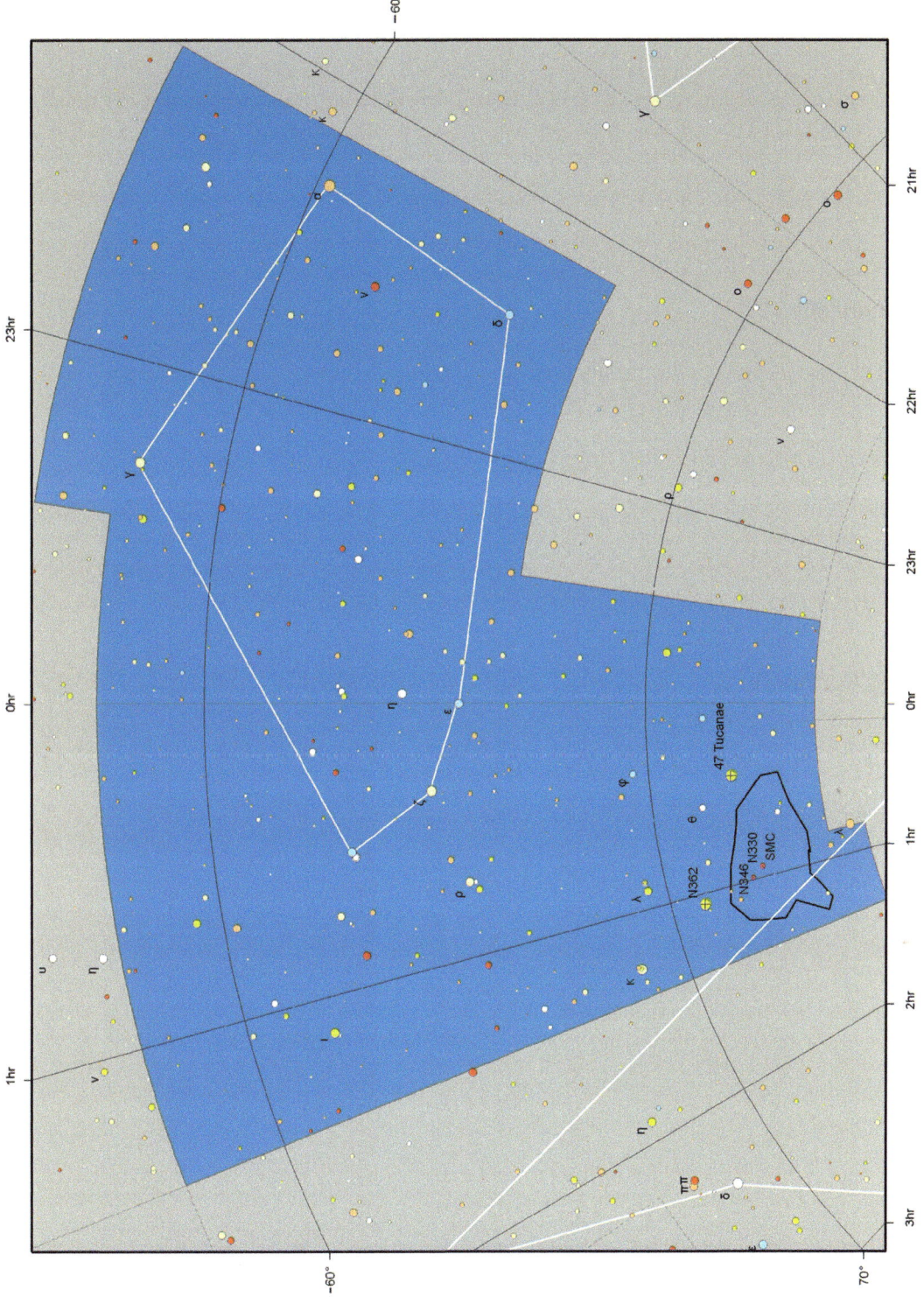

URSA MAJOR

One of the iconic northern hemisphere constellations. It's supposed to represent a large bear. But many see it as a spoon, plough or "big dipper". The handle of the spoon is seen as the bear's tail. The brightest seven stars are all more than magnitude 3.3, making the constellation easy to spot.

The two stars at the end of the spoon are known as the pointers, because a line drawn through them points in the direction of Polaris and the Celestial Pole. They provide a convenient way to find the direction north lies in.

Several of the stars that make up the spoon formed from the same cloud of dust/gas and are travelling through space together. The others are not: in a few thousand years, the shape of the spoon will have changed.

Historically

Ursa Major is an ancient constellation and the spoon shape – which is but part of it – is one of the most easily recognized shapes in the sky. It's seen as depicting a bear facing west, so that the handle of the spoon is the long tail of the bear – which is odd as bears do not have noticeable tails. It does make you wonder if the translations are quite as accurate as historians suppose. Either way, the bear is said to be Callisto, who was turned into one by Artemis as a punishment after she became pregnant following an encounter with the thoroughly vile Zeus. It's not a happy story.

Notable Stars

Double: Zeta Ursae Majoris. 2.4/4.0 and 12 arc minutes – an easy naked eye pairing. Alcor and Mizar. The brighter of the two stars that make up this naked eye double star is, itself, a double. 2.5/4.1 14″.

Double: Nu Ursae Majoris 3.5/9.5 and 7″. Orange and blue.

Double: Xi Ursae Majoris 4.3/4.8 1.8″. Yellow stars.

Deep Sky Objects

M81/M82: A pair of bright galaxies that can be tracked down by sweeping north west of the pointers. M81 is the larger of the two, and is an inclined spiral. It appears as magnitude 7 and spans 25 × 15 arc minutes. M82 is a fainter edge on spiral galaxy undergoing a burst of star formation. Its edge-on nature means that dust in the plane of the galaxy will give it a mottled appearance in modest instruments.

M101: A 7th magnitude face-on spiral galaxy almost 30 arc minutes across, it is most easily spotted with low powers and 150mm aperture. Its light is so spread out that it has a very low surface brightness and can be tough to find. It may even be easier to spot using a finder, rather than your main telescope.

M97: A dim planetary nebula that's at the limit of many observers using a 150mm aperture in suburban skies. The two dark areas that led to its name of The Owl are not easily glimpsed. The nebula is 3.3 arc minutes across.

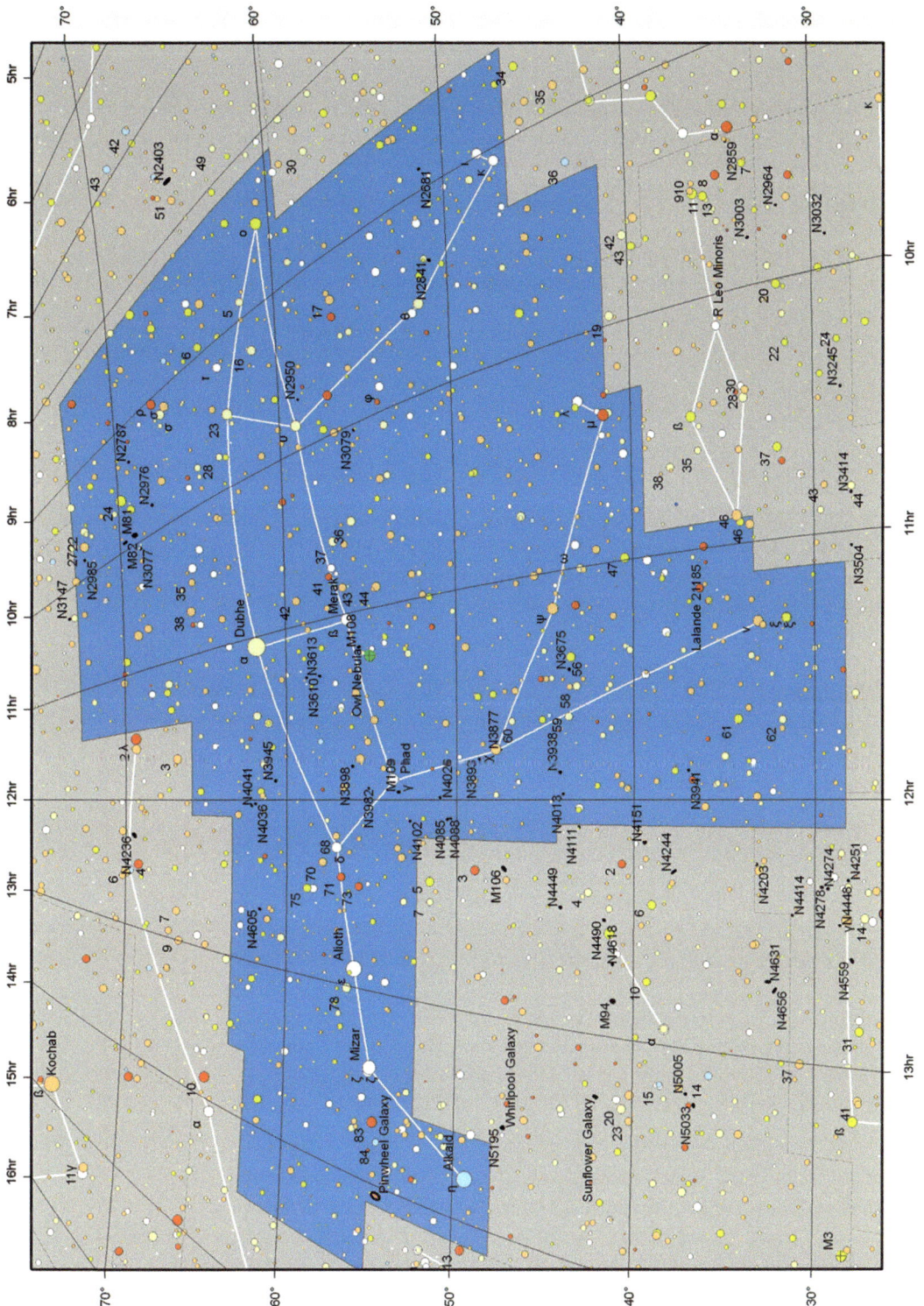

Fig. 27 The inclined spiral M81 is a binocular object lying near to M81 in Ursa Major (Image by Grant Privett)

URSA MINOR

A northern hemisphere constellation – in fact the most northerly- with its brightest star, Polaris, laying within 0.5° of the North Celestial Pole, about which the entire sky appears to rotate. Its shape is very similar to Ursa Major, with Polaris marking the end of the handle of the "Big Dipper". Unfortunately, it's nothing like as easy to spot as Ursa Major, because the stars are a lot dimmer. With two shining at 5th magnitude, this means suburban dwellers may have trouble picking out the full shape.

Polaris is a Cepheid star that has been seen in the past to pulsate. However, in recent years that seems to have diminished, so the star now shines continuously at a near magnitude 2.0.

On 22nd December meteor activity may be spotted. This appears to emerge from the area of the constellation south of the second magnitude star, beta Ursae Majoris, Kocab.

Historically

Ursa Minor is an ancient constellation depicting a small bear. The Greeks reported it, but it is highly likely that they inherited it from previous Mediterranean civilisations, such as the Babylonians. As with Ursa Major, it's a bear shape featuring a prominent tail. It's worth remembering that, at the time of the Greeks, alpha Ursae Minoris (Polaris), was not located particularly near to the North Celestial Pole and, so, didn't have quite the importance we assign to it today. Hang around until 4000 AD and you will see what I mean. Precession of the axis of the Earth makes the pole appear to move relative to the stars, following a leisurely 22,000 year cycle.

Notable Stars

Double: Alpha Ursae Minoris 2.0/9.0 18.4″. White stars.

Double: Gamma Ursae Minoris, 3.1/5. A line of sight naked eye pairing with 11 Ursae Minoris.

Variable: Alpha Ursae Minoris, Polaris, is a giant star and Cepheid variable as discussed earlier. The period is nearly 4 days and the amplitude is now 0.1 magnitudes or less – too small for visual estimation.

Deep Sky Objects

There are no notable deep sky objects in Ursa Minor.

VELA

A southern hemisphere constellation that is best placed for study during March. Its stars appear against the backdrop of a rich part of the Milky Way, with several lovely clusters within its bounds.

The stars delta, kappa, iota and epsilon Velorum are sometimes known as the False Cross and can allegedly be mistaken for the Southern Cross itself. Personally, I have never had trouble telling the two apart – there's no contest – and I suspect few first time visitors to the southern hemisphere would confuse the two.

Historically

Vela was once part of the enormous, but now defunct, Argo Navis constellation. It was separated out by de Lacaille to form a constellation representing the sails of Jason's ship. During the separation process the original star labelling was retained, so that its brightest star is now the first magnitude supergiant, gamma Velorum.

Notable Stars

Double: Gamma Velorum 2.5/4.0 and 41″. There's also an 8th magnitude member at 62″ and a 9th magnitude outlier at 93″. Well worth looking at. Unusual: Gamma Velorum is a Wolf-Rayet star. It's a brilliant blue O type star where enormously powerful solar winds are carrying away the stars outer layers to form a nebula. The star has lost more than two thirds of its mass during its short furious life but, with 9 solar masses remaining, it is still en route to become a supernova. It's the only such star visible to the naked eye.

Double: Delta Velorum 2.0/6.5 and 2.6″. Three other dimmer stars are visible nearby.

Deep Sky Objects

NGC3132: Lying just on the border with Antlia is an oval, 8th magnitude, planetary nebula that appears 45×65 arc seconds across. It has a 10th magnitude central star. It's located roughly 2,000 light years from us.

NGC2670: A 7th magnitude open cluster, roughly 6 arc minutes across. Sparse but with some nice alignments of stars. May appear as a faint glow in 10×80mm binoculars. Look for 20, or so, stars.

IC2391: A naked eye cluster near delta Velorum. A collection of 30 stars within 40 arc minutes, which combine to have a brightness of a 2nd magnitude star. Dominated by 6 stars.

NGC2669: A compact cluster of 20 stars within 8 arc minutes in an area of sky that is lovely when swept with binoculars.

NGC 2547: Another naked eye open cluster. It covers nearly 20 arc minutes.

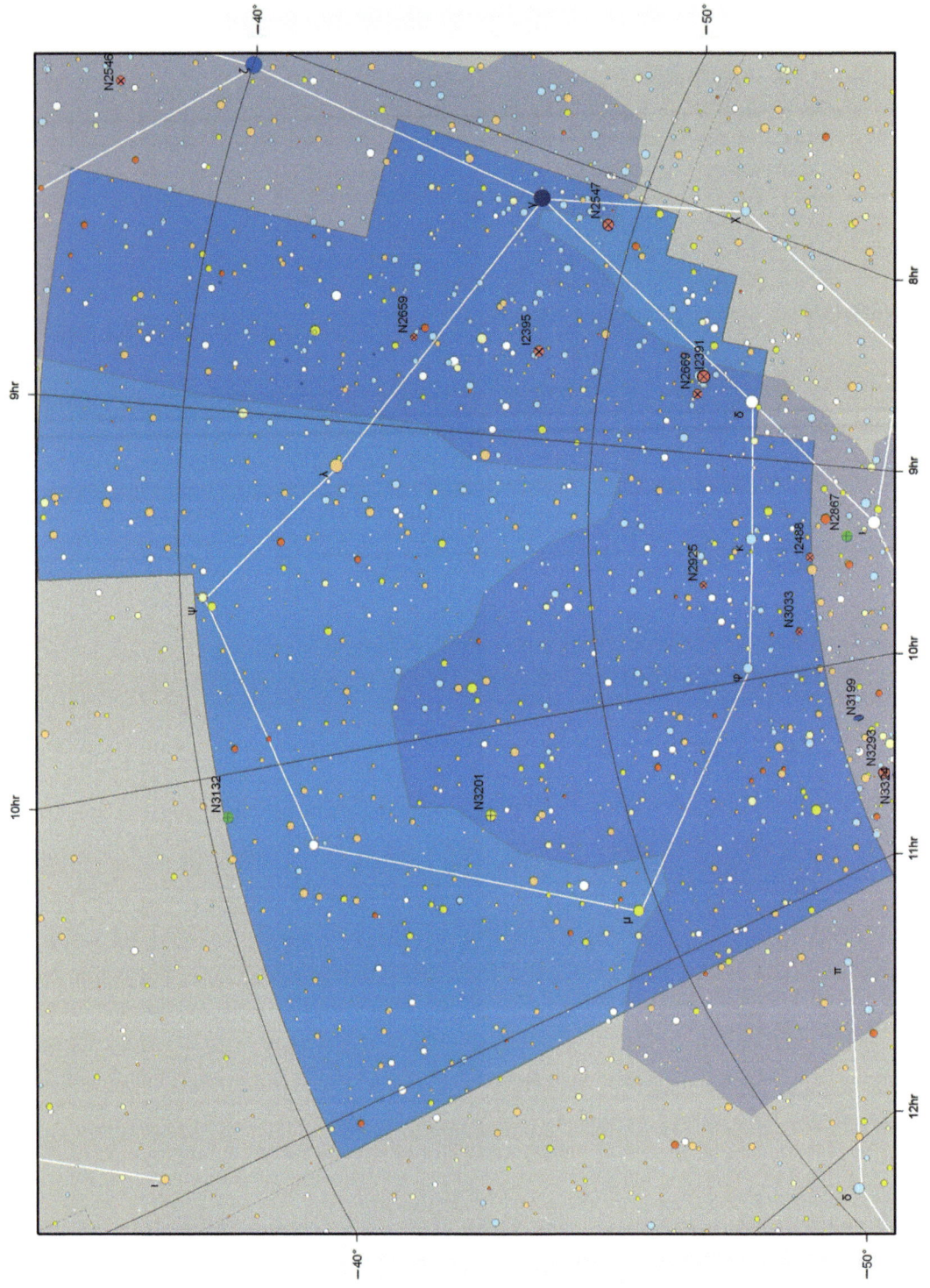

VIRGO

A large constellation, straddling the celestial equator, with an easily remembered Y-shaped outline. It's best placed for observation during the month of May and is dominated by the brilliant 1st magnitude star Spica. The ecliptic passes through the constellation so it is not uncommon for a bright planet to lie in close proximity (in conjunction) with Spica. Occasionally the Moon passes in front of it, which is fun to watch.

Virgo is best known for its galaxy cluster. A vast assemblage of more than 1,800 galaxies, varying from small ultra compact dwarf galaxies (100 million stars), up to the mammoth cannibalistic elliptical galaxy M87 (100 trillion stars). Under a dark sky, with a decent telescope, there are so many notable galaxies it's possible to get confused as to which one you are actually looking at. Under urban skies, galaxies are more difficult to spot, as filters do not do much to make them more apparent. Look for dim, ghostly patches of light.

Historically

Virgo is an ancient constellation depicting a woman. After that, it's very much: which legend do you fancy? There's Demeter (mother of Persphone), Dike (Astraeia) and, perhaps, even the Syrian Goddess Shala. Whichever one you choose, the bright star Spica is generally thought to be an ear of corn held to the lady's ear. No shrinking violet, Shala also carried a serious sword.

Notable Stars

Double: Gamma Virginis (Porrima) 3.5/3.6 and 2.0″. Bright yellow stars.

Double: Theta Virginis 4.5/8.9 and 7″. White and yellow stars.

Quasar: 3C273 lies close by 16 Virginis. It appears as a 12th magnitude blue star, but is in fact 2.4 billion light years away and thus enormously bright.

Variable: R Virginis is a Mira type variable star, varying between 6th and 12th magnitude in a period approximating 140 days.

Deep Sky Objects

M87: A bright and cannibalistic supergiant elliptical galaxy, with a super massive black hole at its core, spewing out a jet of material best seen in images. Appears 6 arc minutes across and 9th magnitude

M49: Another bright elliptical galaxy. 8×6 arc minutes and 9th magnitude.

M104: A classic near edge-on spiral galaxy with a strong central bulge. The Sombrero Hat galaxy shines at 9th magnitude and is 7×3 arc minutes across.

M61: A face-on spiral galaxy with loose arms. Tenth magnitude and 5 arc minutes across. A fascinating target.

NGC5634: A 10th magnitude 4 arc minutes wide globular. Adjacent to an 8th magnitude star.

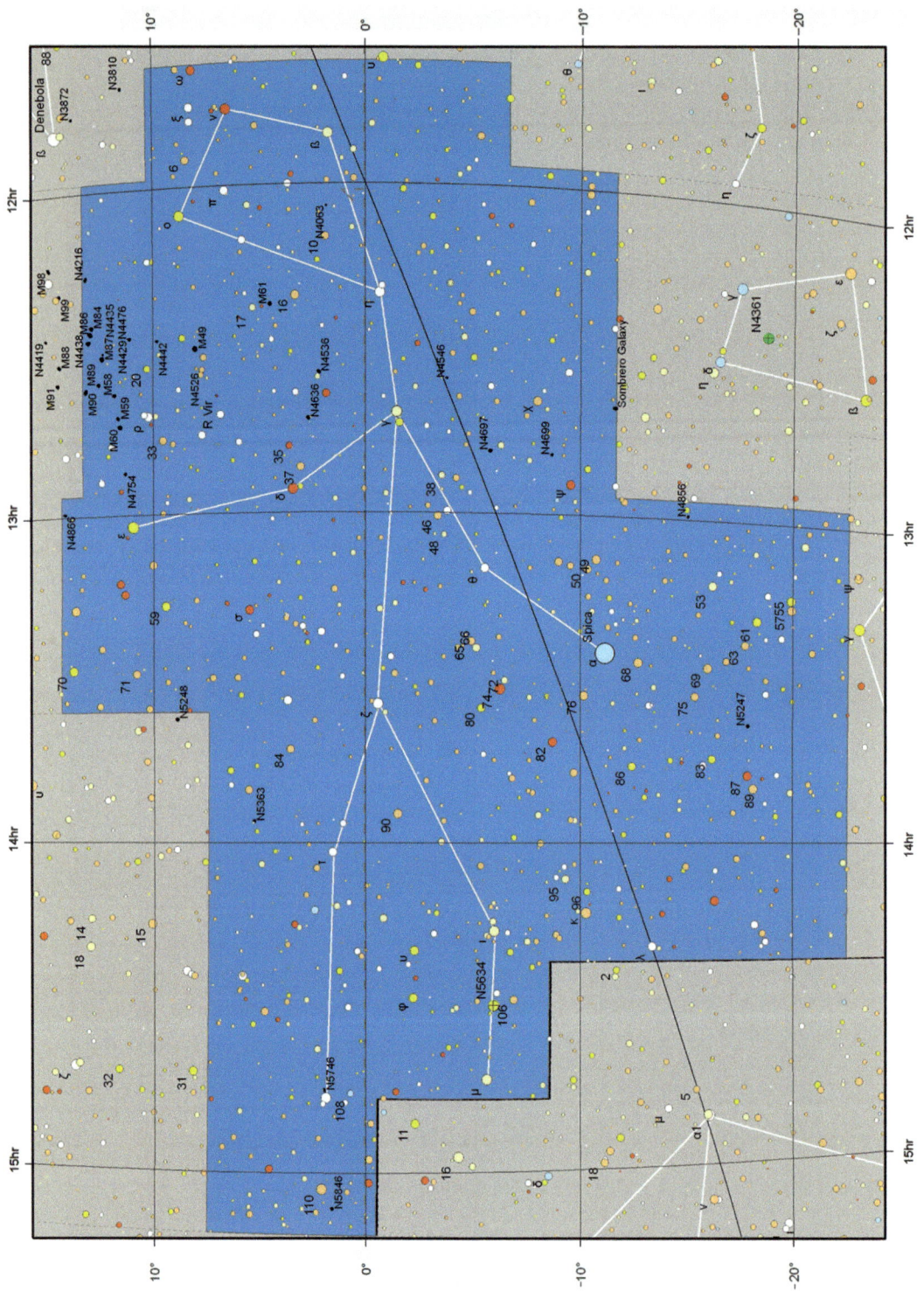

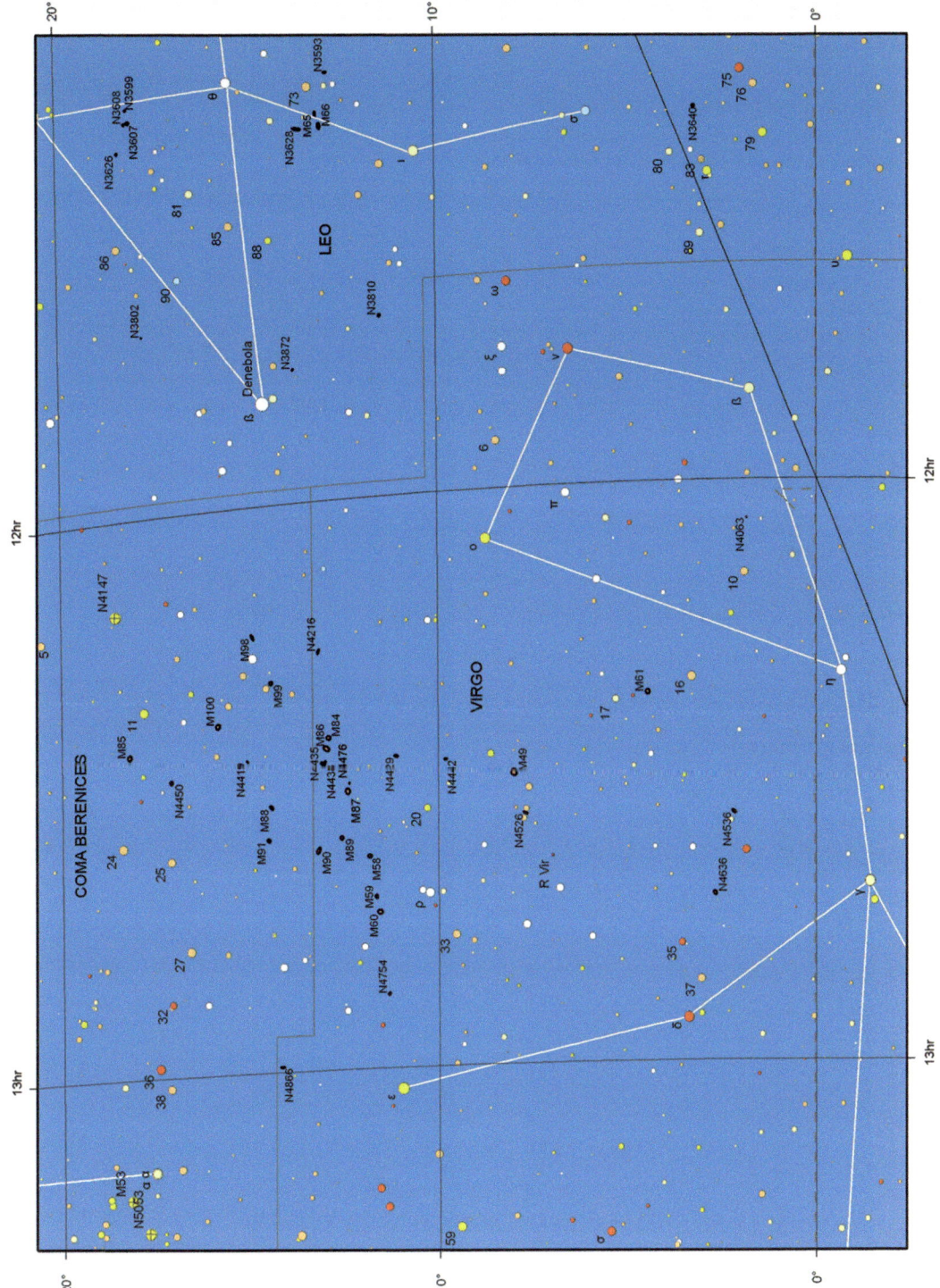

VOLANS

A small and not very bright southern hemisphere constellation; it represents one of the flying fish which captivated early mariners.

Whilst lying in an interesting area of the sky between the Large Magellanic Cloud and the Milky Way, it does not, itself, contain an awful lot that's worth talking about.

It's one of a number of constellations where the brightest star is not denoted alpha. Alpha Volantis, is a star of magnitude 4.0 while beta and gamma are both magnitude 3.8. Look for it in March.

Historically

Volans was first introduced in the star charts of Petrus Plancius during 1598. It was based upon observations undertaken by the Chief Navigator and the First Mate, (Keyser and de Houtman, respectively), of an ill-fated Dutch ship that sailed to the East Indies in 1595. It was one of a number of constellations introduced by Plancius.

Notable Stars

Double: Gamma Volantis 4.0/5.5 and 14″. Yellow stars. Pretty.

Double: Epsilon Volantis 4.5/8.0 and 6″. Blue and white stars.

Deep Sky Objects

There are no notable deep sky objects in Volans.

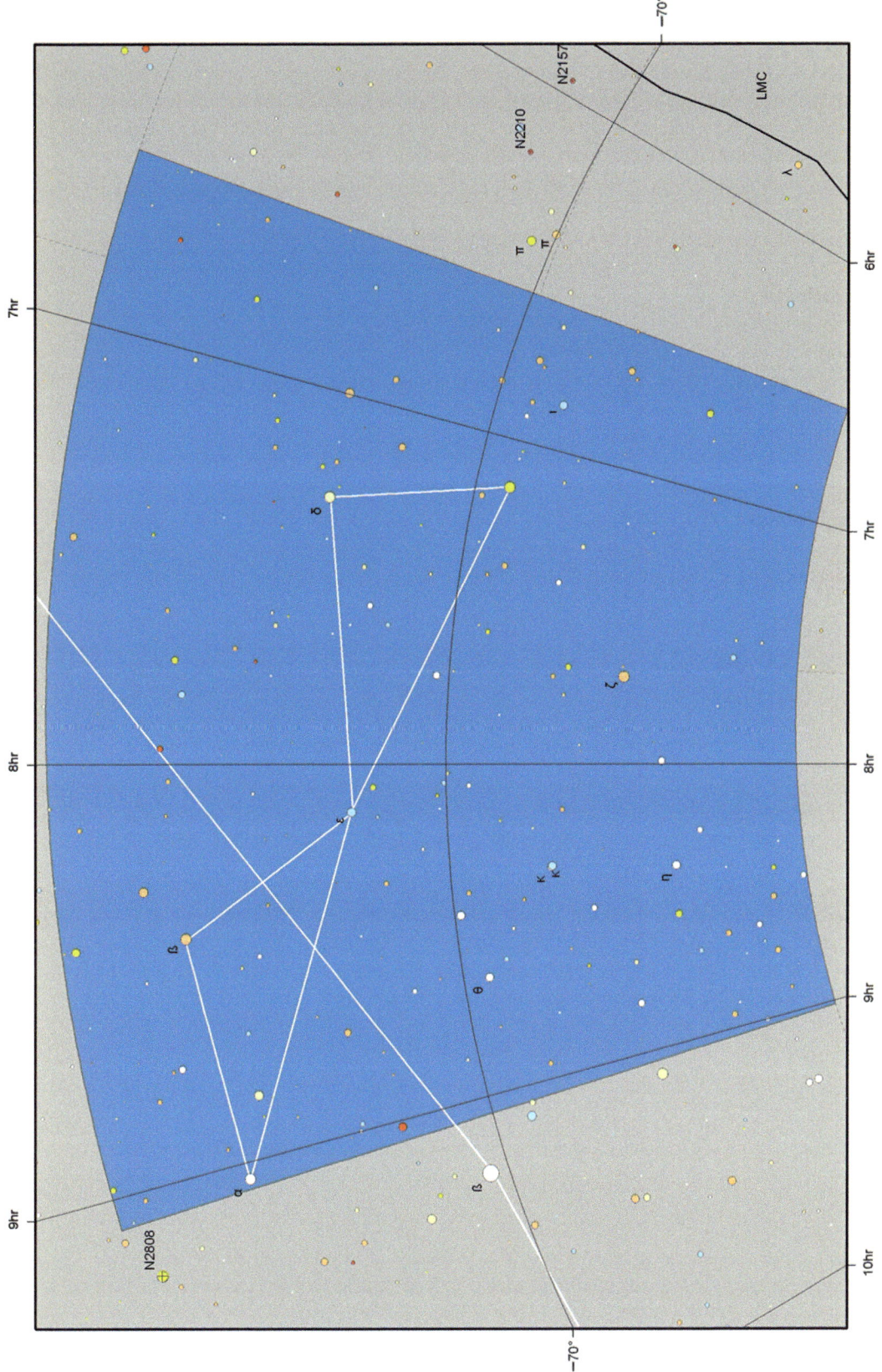

VULPECULA

A small constellation with a Milky Way backdrop, which inevitably leads to rich star fields – ideal for sweeping with a small telescope, or binoculars. It's not an especially interesting shape and none of its stars are above 4th magnitude, but it is, at least, easy to find; nestling between Lyra and Sagitta – which provides a much more obvious and memorable shape.

While there are relatively few notable stars, the location does boast a couple of worthwhile deep sky objects.

It is best observed during the month of September.

Historically

Vulpecula represents a fox. Vulpecula is a surprisingly recent addition, having been created by Johannes Hevelius in 1687. Looking at some of the early charts makes one wonder how often that he had seen a live fox, as the representations look more like a deformed dog.

Notable Stars

No double stars.

Deep Sky Objects

M27: The Dumbbell nebula. A bright planetary nebula, easily spotted in 10×50mm binoculars. It is oval in small scopes, with 150mm instruments marking the point where the two lobes become noticeable.

Coathanger: A small grouping of 10 or so stars which looks very much like, wait for it, a coathanger. The members are between magnitudes 5 and 7 and so easily spotted in binoculars and are well shown in a low power wide field view with a small aperture instrument. An 80mm refractor is ideal. Also known as Brochi's cluster and Cr399.

NGC6940: A binocular cluster of 7th magnitude. It contains many faint stars and so doesn't resolve particularly easily under urban skies. More than 30 arc minutes across – the width of the full Moon – with a bright central star. A good test for sky clarity. If you can see it with 10×50 binoculars, the sky conditions are usable for deep sky observing.

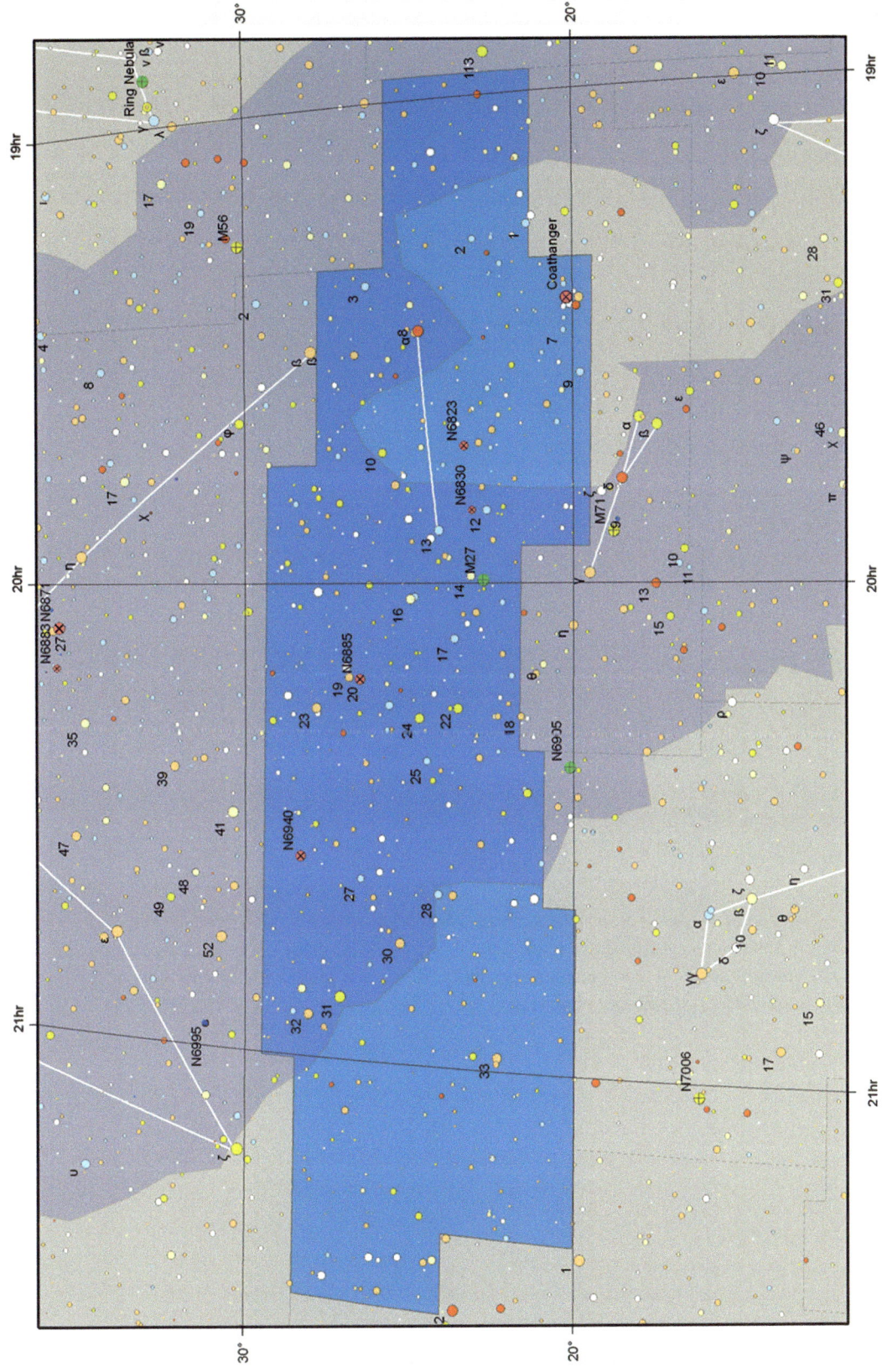

Fig. 28 The bright planetary nebula M27 in Vulpecula – visible in binoculars (Image by Bill Snyder)

Fig. 29 The Coathanger asterism in Vulpecula – well suited to binocular observing (Image by Greg Parker)

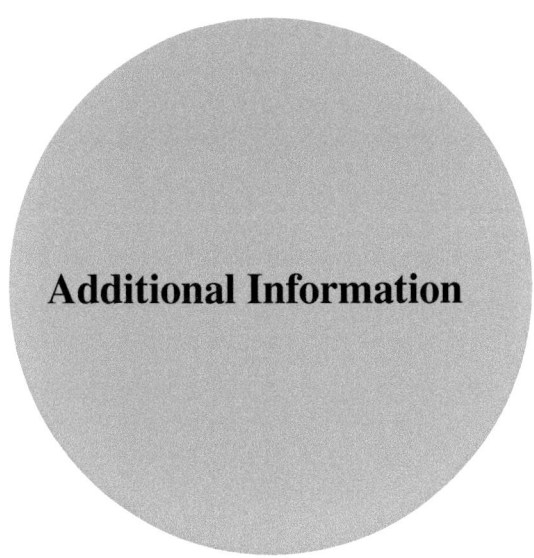

Additional Information

SKY LORE

Those curious to hear more about the origins of the constellations should read Ian Ridpath's superb book *Star Tales* from Lutterworth Press. The author even credits illustration to John Flamsteed and Johann Bode as co-authors. It covers the subject in far greater depth than we have space to provide here and is an engaging read.

Accounts of sky mythology from non-European cultures are covered by Robert Burnham in his three volume *Celestial Handbook* published by Dover Books.

PLANETS

The Planets and the Moon are great objects to observe and can be followed, even from the centre of a city. Books on the subject include:

Gerald North – *Observing the Solar System: The Modern Astronomers Guide*, Cambridge University Press.

Patrick Moore – *A New Guide to the Planets*, Sidgwick and Jackson. A bit dated but, as always, an entertaining read.

Martin Mobberley – *Lunar and Planetary Webcam User's Guide*, Springer.

DEEP SKY OBJECTS

If this book has piqued an interest in deep sky objects then you might like to try some of the books below:

Neil Bone and Will Tirion – *Deep Sky Observers Guide*, Firefly.

Grant Privett – *The Deep Sky Observers Year*, Springer.

Sue French – *Deep Sky Wonders*, *Sky and Telescope*, Firefly.

METEORS

A great wealth of information regarding meteor showers can be found at the websites of the British Astronomical Association (http://britastro.org/baa/) and also the International Meteor Organization (http://www.imo.net). There's lots of scope for helping to observe and measure the less active showers.

COMETS

Most comets are rather dim and unimpressive, but a surprising number are within range of smaller telescopes and binoculars. Follow the latest events in magazines like *Astronomy Now* http://www.astronomynow.com and *Sky and Telescope* http://www.skyandtelescope.com

WEB BASED SUPPORT

There's a lot of information on the web, but it is of highly variable quality, which can make it a daunting task to find the good stuff. Typing "observing the stars" into Google today (January 2013) threw up 16.6 million hits. That deluge of data means that when you know little about the subject it is difficult to distinguish between the expert, the opinionated and the fool. For example, its impressive how many websites depicting themselves as experts have taken their content, lock, stock and barrel (complete with errors) from Wikipedia. While that use may be perfectly legal, if you are not going to write your own stuff why not just say, "Look in Wikipedia".

One way to avoid the pitfalls is to start with a general introductory text – the late Patrick Moore wrote many – and via a subscription to a magazine like *Astronomy Now* where an editor is there to provide a quality control. You won't assimilate it all in one go – don't try – but over a few months you will learn a lot.

Membership of an organization of like-minded enthusiasts such as a local astronomical society/club, for instance the British Astronomical Association, will also help. Their newsletters are informative and getting together once a month to hear what other people are up to and what is visible in the month ahead keeps the enthusiasm nicely stoked. Societies often run observing sessions where you can turn up with or without a telescope and see some of the more impressive deep sky sights the sky has to offer, thereby allowing you to gain experience of learning to "see" faint dim objects without the hassle of having to check you have the telescope pointing in the right direction

It will also provide the opportunity to hear first-hand accounts of telescopes and mounts from people that actually use them. Often, discussing telescopes will lead on to a discussion of what instruments to avoid, and this can provide some of the strongest steers and overcome the exaggerated claims and hyperbole of manufacturers.

Also, people forget the social side of astronomy: it's not clear every night and sharing plans and observing sessions can be great fun.

Appendix

THE IMAGE CONTRIBUTORS

While reading or browsing this book you will come across a number of images of deep sky objects. These were all taken by amateur astronomers. Since the advent of CCD cameras, the quality of the images taken by amateurs has increased rapidly, to the point where the images taken in a suburban location can now equal those taken by professional astronomers during the 1980s. It is now quite common for amateur astronomers to image objects that cannot even be seen on the Palomar Schmidt Survey plates.

Below we list those generous individuals who supplied the images and give a little information about their activities. Their websites are well worth visiting. Please remember that when viewing objects at an eyepiece, your eyes are unlikely to detect any color – the objects are simply too dim and appear grey.

Bill Snyder http://billsnyderastrophotography.com/
Bill is a resident of Pittsburgh, Pennsylvania in the United States where he images using a TMB 130 mm refractor combined with an Apogee CCD at his light polluted home observatory. In addition, he operates a Planewave 17″ telescope at a remote observatory in the Sierra mountains.

The Omega Centauri image used herein was created by Bill using data collected by the Australian astronomer **Martin Pugh**.

Chris Picking http://www.starrynightphotos.com
Chris is an astronomer based in a dark area of New Zealand's north island. He images using a Canon DSLR and *hand-guided* 6 min exposures on a Vixen GPDX telescope. He also images using

an 8″ Newtonian reflector in his hand crafted observatory. His images of the southern sky make me want to emigrate to New Zealand – or at least visit it.

David Ratledge http://www.deep-sky.co.uk/

David lives in Lancashire, just 15 miles from Manchester and so observes from one of the UK's wettest counties. Despite this, he creates a wealth of images using equipment from RCOS, Takahashi, Apogee and Canon. He successfully overcomes the dreadful conditions, and images objects as diverse as hugely remote brilliant quasars and our Sun. Particularly impressive is an image of a $Z > 5$ quasar.

Prof. Greg Parker http://www.newforestobservatory.com

Greg is a deep–sky imager based in the New Forest area of the United Kingdom. He images with a Celestron Nexstar 11 GPS via a Hyperstar attachment using Starlight Xpress cameras. Most of his images are generated in collaboration with Noel Carboni, an image processor from Florida, USA. To collect data more quickly, he is currently creating a 3-tube assembly survey instrument – Mini-WASP – which sits atop a Paramount ME.

Index

General

Almagest, 68, 89, 118, 183
Aphrodite, 160
Apollo, 72, 104
Argo Navis, 1, 46, 165, 168, 203
Artemis, 177, 198
Babylon, 1, 3, 15, 18, 104, 120, 122, 145, 163, 174,
    192, 201
Bayer, 196
Bode, Johann, 93, 133, 138, 215
Boyle, Robert, 11
Callisto, 198
Catalogue, 112, 122, 138, 170
Chiron, 52
Declination, 1–4, 40, 44, 60, 87, 142, 165
De Houtman, Frederick, 13, 60, 97, 108, 208
De Lacaille, Nicolas Louis, 11, 28, 46, 60, 62, 102, 129,
    138, 140, 158, 168, 170, 179, 190, 203
Egypt, 40, 76, 145
Eos, 15
Equinox, 22, 76
Eros, 160
Ethiopia, 50, 55
Eudoxus, 18
Europa, 187
Eurydice, 126
Flatfoot, Peter, 84
Ganymede, 15
Greek, 1, 3, 5, 8, 15, 18, 20, 22, 24, 44, 55, 66, 68, 76,
    84, 95, 101, 104, 120, 145, 150, 160, 174,
    187, 192, 194, 201
Gutenberg, 133
Hadley, John, 140
Hera, 33, 87
Heracles, 99
Hercules, 33, 87, 99, 101, 104, 112, 126, 172

Hevelius, Johannes, 36, 110, 116, 124, 181, 185, 210
Homer, 26
Huyghens, Christian, 102
International Astronomical Union (IAU), 1
Janssen, Zacharius, 131
Jason, 22, 46, 165, 203
Kepler, 84
Keyser, Pieter, 13, 60, 84, 97, 108, 136, 208
Kirch, Gottfried, 91
Leda, 79, 95
Medusa, 150, 153
Mercury, 153
Merope, 145
Nereid, 7, 50
Nergal, 174
North celestial pole (NCP), 3, 30, 201
Olympus, 15
Orpheus, 79, 126
Pan, 44, 160
Papin, Denis, 11
Plancius, Petrus, 13, 30, 60, 64, 84, 106, 108, 133, 136,
    148, 156, 194, 196, 208
Planets, 3, 4, 22, 68, 160, 177, 187, 205, 215
Platevoet, Pieter, 84
Poseidon, 7, 82, 150, 153
Prometheus, 18, 172
Proper motion, 2
Ptolemy, 18, 24, 26, 33, 55, 66, 68, 70, 76, 89, 118, 122,
    177, 183
Right ascension, 1, 3, 4, 87, 104
Roman, 1, 18, 76, 95, 99, 104, 145, 183, 192, 194
Royal Society, 36, 110
Shala, 205
South celestial pole (SCP), 3, 13, 60, 106, 140, 196
Sumerian, 3, 44, 177, 194
Summer triangle, 2, 18

G. Privett and K. Jones, *The Constellation Observing Atlas*, The Patrick Moore Practical Astronomy Series,
DOI 10.1007/978-1-4614-7648-1, © Springer Science+Business Media New York 2013